METHODS OF CHARACTERIZATION OF SEWAGE SLUDGE

Commission of the European Communities

METHODS OF CHARACTERIZATION OF SEWAGE SLUDGE

Proceedings of a Workshop held in Dublin, July 6, 1983

Edited by

T. J. CASEY
University College, Dublin, Ireland

P. L'HERMITE
Commission of the European Communities, Brussels, Belgium

and

P. J. NEWMAN
Water Research Centre, London, United Kingdom

D. REIDEL PUBLISHING COMPANY

A MEMBER OF THE KLUWER ACADEMIC PUBLISHERS GROUP

DORDRECHT / BOSTON / LANCASTER

Library of Congress Cataloging in Publication Data
Main entry under title:

Methods of characterization of sewage sludge.

At head of title: Commission of the European Communities.
Bibliography: p.
Includes index.
1. Sewage sludge–Characterization–Congresses. I. Casey, T. J. (Thomas Joseph), 1933- . II. L'Hermite, P. (Pierre), 1936- . III. Newman, P. J. IV. Commission of the European Communities. V. Commission of the European Communities. Directorate-General for Science, Research and Development.
TD767.4.M48 1984 628.3'6 84-8285
ISBN 90-277-1782-6

CIP

Organization of the Workshop by
Commission of the European Communities
Directorate-General Science, Research and Development, Brussels

Publication arrangements by
Commission of the European Communities
Directorate-General Information Market and Innovation, Luxembourg

EUR 9171

Published by D. Reidel Publishing Company
P.O. Box 17, 3300 AA Dordrecht, Holland

Sold and distributed in the U.S.A. and Canada
by Kluwer Academic Publishers,
190 Old Derby Street, Hingham, MA 02043, U.S.A.

In all other countries, sold and distributed
by Kluwer Academic Publishers Group,
P.O. Box 322, 3300 AH Dordrecht, Holland

Printed in The Netherlands

PREFACE

A.M. BRUCE
Water Research Centre, Stevenage, United Kingdom

Sewage sludges from different sources can exhibit wide differences in their physical, chemical and biological characteristics. In practical terms, this means that their behaviour in the various processing operations (pumping, thickening, stabilisation, conditioning, dewatering) and their characteristics after processing can vary in an unpredictable way. To arrive at the most appropriate choise, specification and method of operation of a sludge processing plant, it is usually desirable to characterise the particular sludge in respect of the relevant processing parameters using laboratory or pilot-scale tests.

The development of tests for characterising various sludge parameters has been proceeding for some years in a number of countries. From the commencement of the COST 68 Concerted Action on Sewage Sludge, there has been an effort to coordinate this area of research and in particular to identify new parameters of importance and to standardise definitions and test methods. The characterisation tests relating to sludge processing have been the special responsibility of the COST 68 Working Party 1, while chemical and biological characteristics relating to problems of environmental pollution have been the concern of the Working Parties 2 and 3 respectively.

In order to review the state of development of laboratory and pilotscale characterisation tests relating to sludge processing, and to identify further research needs in this area, Working Party 1, with the support of the Commission of the European Communities, organised a Workshop in Dublin on 6 July 1983. The papers presented cover most of the more recent work within the COST 68 participating countries. This volume contains the proceedings of the Workshop together with some brief conclusions contributed by members of Working Party 1. It is hoped that the information provided will be valuable both to plant designers and operators and to those involved in research on sludge processing techniques.

C O N T E N T S

CONCLUSION

SESSION I

Chairman: Mr. A.M. BRUCE

Rapporteurs: Prof. T.J. CASEY
Dr. P.J. NEWMAN

SELECTION AND USE OF POLYELECTROLYTES IN DEWATERING OF SLUDGE WITH BELT PRESSES

B.A. HEIDE and R. KAMPF
TNO Research Institute for Environmental Hygiene,
P.O. Box 214, 2600 AE DELFT, The Netherlands.

Summary

A new procedure is presented for the characterization of sludge for dewatering with belt presses and interpretation of practical performance, for selection of suitable polyelectrolytes and optimum use of the selected polyelectrolyt . The procedure evolved from investigation on laboratory scale and in practice in which a large of polyelectrolytes and different belt presses were involved.
The procedure comprises a combination of conditioning at laboratory scale and the application of a newly developed, very simple dewatering test (MFT test) which furnishes an insight into the rate of dewatering as well as in the attainable dry solids content.
The applicability of the procedure in practice was verified at more than ten sewage treatment plants where aerobically stabilized and digested sludges are dewatered with belt presses.

1. INTRODUCTION

In dewatering of sludge with a belt press several factors should be considered. These factors include characterization of sludge for this type of dewatering equipment, selection of suitable polyelectrolytes and optimum use of the selected polyelectrolyt.

By order of the Foundation for Applied Research on Sewage Treatment (STORA) the TNO Research Institute for Environmental Hygiene has carried out investigations on the optimalization of dewatering of sludge with belt presses in which the above-mentioned factors have been studied. Besides reports and publications a manual on the selection and use of polyelectrolytes in sludge dewatering with belt presses has been set up [lit.1,2,3]. This publication comprises a short summary in which especially the applicability of the developed procedure is illustrated.

2. LABORATORY METHODS

For a detailed description of the various tests, including the determination of the excess polyelectrolyt in the filtrate and the turbidity of the filtrate, we refer to the literature mentioned. Here only a new dewatering test the MFT test (suction time and % dry solids) and the standard conditioning test are discussed.

2.1. MFT test

The Modified Filtration Test (MFT) is in essence a modification of the Büchner filtration test. The MFT test was developed by TNO with the aim to obtain in one simple test insight both into the rate of dewatering and the attainable dry solids content after dewatering with belt presses [lit.4].

Method

The test comprises a vacuum filtration of conditioned sludge which is carried out with a 7 cm diameter Büchner Funnel (e.g.SS 589^2 Weisband filter paper) at 0.5 bar during 10 minutes. Cracking of the filter cake or drawing in of air by the edge of the cake is prevented in the following way. After filtration has begun a plastic foil is laid loosely and with ample space over the Büchner funnel. On the foil a layer of water (2 cm thick) is brought causing the plastic foil to fit closely at the upper side of the sludge and the inside of the funnel. After 10 minutes of pressing the test is ended and the foil with the water is removed.

Rate of dewatering

As a measure for the rate of dewatering the suction time is used. This is the time in which 60 ml of filtrate is obtained starting from 120 ml of conditioned sludge. If desired the volume of filtrate can be measured as a function of time and thus the specific resistence at 0.5 bar can be calculated. The suction time agrees well with the specific resistance and the capillary suction time (CST). However, it appeared to be more sensitive for small changes in dewaterability of sludge then specific resistance or CST.

Final dry solids content

In older literature on sludge processing the cracking time is mentioned; this is the time the cake needs to crack during vacuum filtration. This method is no longer applied because of its less satisfactory accurary. In this respect, the suction time is more practicable and proceeding the filtration in the afore-mentioned way the filtercake is pressed out. After 10 minutes of filtering/pressing the dry solids content of the cake is determined. This dry solids content can be related with the attainable dry solids content of the sludge when dewatered on belt presses.

2.2. Standard conditioning test

The standard conditioning test is a version of the "Recommended standard method for assessing the conditionability of a municipal sewage sludge" from the Final Report, Cost Project No.68, Sewage Sludge Processing [lit.5]. The test is carried out in combination with the MFT test.

In the test a 250 ml (low format) beaker is used in which the sludge is stirred by a standard stirrer at 1000 revolutions per minute (see fig.1). The distance from the stirring rod to the bottom is 15 mm. A 20 ml piston pipette is also required. In the procedure to be followed the final volume of the sludge suspension is always the same (120 ml), regardless the amount of polyelectrolyt added.

From the series of stirring or mixing times (1, 2, 5, 10, 20, 35, 50, 80 and 110s) for sludge and flocculant, at least four are selected, after which various dewatering parameters are determined. The optimum stirring time depends on the type of sludge, the type of polyelectrolyt and the dosage of polyelectrolyt. (The optimum stirring time is that time at which the highest dry solids content is obtained with the MFT test).

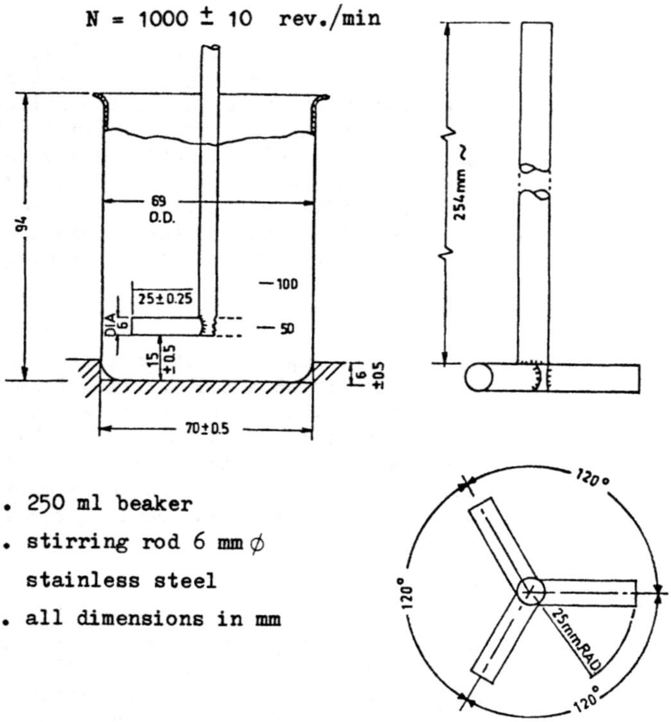

Fig.1: Apparatus for the standard conditioning test (from [lit.5])

The initial quantity is 100 ml of sludge. The concentration of the flocculant solution is thus chosen that the volume of the dosage is always 20 ml. The final volume is always 120 ml.

For characterization of sludge Praestol 444K (as in [lit.5]) is used as standard polyelectrolyt at dosages of 2.5-5-7.5-10 g per kg of dry solids (see ch.3).

For selection of polyelectrolytes the same procedure is followed for other flocculants (see ch.4).

For check on mixing of polyelectrolyt and sludge in practice technically conditioned sludge is used in this test (120 ml) and only additional mixing or stirring is carried out (see ch.5).

3. CHARACTERIZATION OF SLUDGE AND PREDICTION OF PRACTICAL RESULTS

If it is considered to use a belt press for dewatering of a given sludge first characterization of the sludge should be carried out. The relation between sludge characterization on laboratory scale and attainable results in practice with various types of belt presses and polyelectrolytes is very important.

Changes in the wastewater treatment process, but also seasonal influences or changes in the composition of the influent often cause a change in sludge properties. In such cases it is desirable to fix the sludge properties by means of characterization.

Also if there are large changes in operating results of belt presses characterization of the sludge is necessary to examine whether there are large changes in the initial sludge properties or whether there are other causes for this.

3.1. Method

The standard conditioning test is carried out with Praestol 444 K and the percentage dry solids with the MFT test is determined (ch.2). For each dosage of polyelectrolyt the highest MFT-% d.s. at the optimum stirring time is selected and represented graphically as in figure 2. Table 1 gives the results obtained in practice with the same sludges with various types of belt presses and polyelectrolytes.

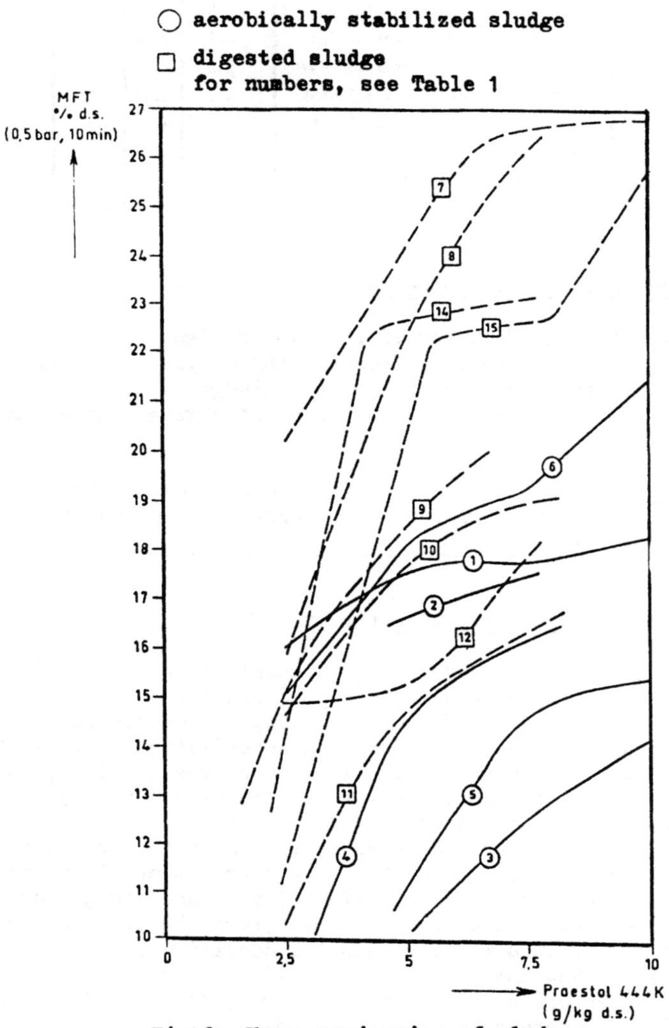

Fig.2. Characterization of sludges

Nr.	Type of sludge	Type of beltpress	Type of p.e.	g p.e./ kg d.s.	% d.s.	Re-marks*
1	aerobic stab.	Klein 15/3 (flat)	Praestol 444K	5.5	12.8	1
1	aerobic stab.	Klein 15/3 (flat)	Praestol 423K	5.5	11.2	2
1	aerobic stab.	Klein 15/3 (flat)	Praestol A751	5.5	12.8	1
1	aerobic stab.	Klein 15/3 (flat)	Superfloc C454	5.5	12.2	1
2	aerobic stab.	Klein S 15	Zetag 63	4	17	2
3	prim./aerobic	Bellmer	Praestol 436K	10	16	3
4	aerobic stab.	Bellmer	Sedipur CF900	4.5	16.5	1
5	aerobic stab.	Klein 15/3 (flat)	Praestol K270	3	11	1
6	aerobic stab.	Willmes	Sedipur CF900	3.5	19	1
7	digested	Klein S 25	Zetag 63	3.5	27.5	2
8	digested	Klein S 25	Praestol 421K	4.5	24.4	4
9	digested	Bellmer	Praestol 433K	3.2	21.2	1
10	digested	Bellmer	Superfloc C470	3.5	19.6	2
11	digested	Klein 10/3 (flat)	Zet.63/P 444K	4.9	12.5	2
12	digested	Bellmer	Praestol 436K	3.6	17.5	3
14	digested	Passavant	Praestol 521K	3.3	18.2	2
15	digested	Klein 15/3 (flat)	Zetag 63	2.5	18.8	1

*** Evalutation based on laboratory results**

1) good polyelectrolyt, good mixing of p.e. and sludge
2) other p.e.'s give better results, good mixing of p.e. and sludge
3) good p.e., insufficient mixing of p.e. and sludge
4) other p.e.'s give better results, insufficient mixing of p.e. and sludge.

Table 1. Results in actual practice related to the characterization of sludges (see curves in fig.2)

3.2. Interpretation of characterization results

The data stated in table 1 and in fig. 2 enable us to evaluate the characterization to attainable results in actual practice.

The characterization results found when using the mentioned procedure are represented graphically in fig.2. This characterization curve is then compared with the other numbered characterization curves in fig.2. The number of the lines refers to table 1 where results from actual practice with corresponding sludges are stated.

Table 1 also states whether according to the laboratory research the best polyelectrolyt (p.e.) was used on the sewage treatment plants concerned in the period considered and/or the mixing of p.e. and sludge was sufficient in the practice. Therefore the table gives the minimum dry solids contents to be attained for the relative conditions.

Table 1 and fig.2 show that higher dry solids contents are attained if non-flat belt presses are used (Bellmer, Willmes, Klein-S, Passavant) and the characterization curves are higher in the diagram.

Fig.2 shows that increasing of the p.e. dosage in nearly all cases leads to higher dry solids contents. In practice, especially the range from 2.5 to 5 g of p.e./kg of dry solids appears important. The range should be given special attention when interpreting the characterization curves.

EXAMPLES

Some examples of the interpretation of characterization results are given below.

- Sludge Nr.1 in fig.2 or sludge with corresponding dewatering properties (characterization curve).

  According to table 1:

  . dry solids content using a flat belt presses: 12-13%;

  . dry solids content expected using other presses: >17-19%
  (compare sludge Nr.2 - Klein S: 17%; and
  Nr.6 - Willmes: 19%).

- Sludge Nr.5 in fig.2 or sludge with corresponding dewatering properties (characterization curve)

  According to table 1:

  . dry solids content using a flat belt press: about 11%;

  . dry solids content expected with other presses: >16%
  (compare sludge Nr.4 - Bellmer: 16.5%; 4.5 g p.e./kg d.s.
  and Nr.3 - Bellmer: 16%; 10 g p.e./kg d.s.)

- Sludge Nr.14 in fig.2 or sludge with corresponding dewatering properties (characterization curve).

  According to table 1:

  . dry solids content using a Passavant belt press: 18.2%;

  . dry solids content with a higher p.e. dosage could be 22% at a minimum since other p.e.'s give better results.
  (compare sludge Nr.15 - flat press: 18.8% and
  sludge Nr. 8 - Klein-S : 24.4%).

- Comparison of sludges Nrs.3 and 4

  According to table 1:

  . both sludges are dewatered with a Bellmer press;

  . the bad properties of sludge Nr.3 (see fig.2) requires a very high p.e. dosage to attain also a dry solids content of 16%.

4. SELECTION OF POLYELECTROLYTES

Polyelectrolytes can be selected by several criteria such as dewaterability, cost price, manageableness (dispersion, dissolution) etc. Here only the dewaterability is taken as a criterium. The polyelectrolytes should preferably be selected from three or more different ones. Usually per polyelectrolyt selection research is carried out at three dosages of polyelectrolyt.

4.1. Method

The standard conditioning test is carried out with each polyelectrolyt followed by the MFT test. The dry solids content with the MFT test at the optimum stirring time is determined (fig.3). Measurement of the suction time as part of the MFT test gives supplementary information on the rate of dewatering and is useful for the choice of stirring time during conditioning at a laboratory scale.

4.2. Selection criteria

The selection is made with a number of polyelectrolytes; for each dosage of p.e. examined, selection takes place by the criteria mentioned below.

Selection criteria

If p.e. gives >90% of maximum d.s. content obtained with MFT test
→ p.e. is suitable.
If p.e. gives 80-90% of maximum d.s. content obtained with MFT test
→ p.e. is moderately suitable
If p.e. gives <80% of maximum d.s. content obtained with MFT test
→ p.e. is unsuitable.

The maximum dry solids content obtained with the MFT test gives the highest value obtained for a given sludge at one given p.e. dosage. The type of p.e. rendering the highest value is qualified suitable, as well as the polyelectrolytes giving at the corresponding p.e. dosage a dry solids content of >90% of the highest value.

A refining of the selection of the polyelectrolytes in the category of suitable can be realized on the basis of the suction time. A short suction time means a rapid dewatering. In spite of a high d.s. content with the MFT test a p.e. may be less suitable owing to a long suction time. When dewatering with a belt press this may lead to problems in the pre-dewatering zone.

Besides, the influence of the stirring intensity on the dewatering properties of the conditioned sludge should also be taken into consideration (see fig.3). The use of good polyelectrolytes which are highly sensitive to the stirring intensity when used in actual practice requires more attention in operation than less sensitive polyelectrolytes.

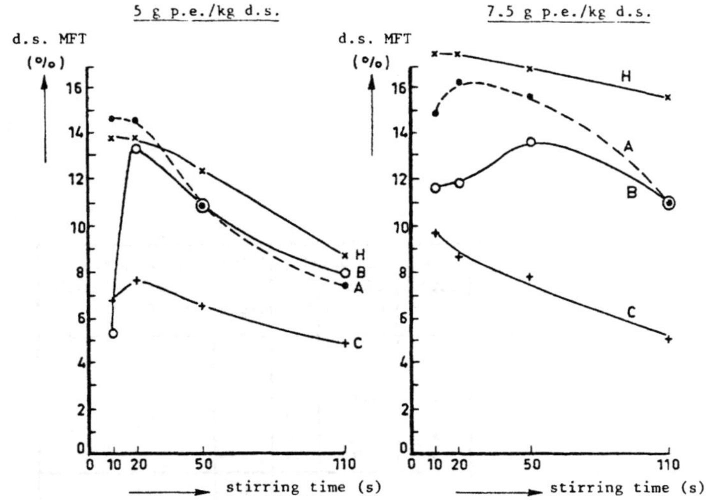

Fig.3. MFT test (% d.s.) of sludge after laboratory conditioning at different mixing times and dosages of polyelectrolytes (for code p.e.'s see table 2).

EXAMPLES

The polyelectrolytes used are stated in table 2.

Polyelectrolyt	% Pos.charged groups	Code	Polyelectrolyt	% Pos.charged groups	Code
Praestol 444 K	100	A	Praestol K 270	80	K
Praestol 423 K	50	B	Praestol 436 K	90	L
Praestol 411 K	25	C	Praestol 421 K	40	M
Praestol A 751	75	D	Praestol 521 K	40	N
Hercofloc 859	high	F	Superfloc C 470	?	O
Superfloc C 454	70	E			
Zetag 63	?	G_1	Mixture of Zet.63 + P.444 K (3:1)	?	Z
Praestol 433 K	80	G_2			
Sedipur CF 900	high	H			

Table 2. Polyelectrolytes used and their codes

The results of the selection on laboratory scale and the comparison with the applied polyelectrolytes in practice is given in table 3.

From the table we see that at the time of the investigation, in eight cases a suited polyelectrolyt, in three cases a less suited polyelectrolyt and in one case an unsuited one was used. This is no absolute evaluation, but it concerns the combination of polyelectrolyt and sludge. Besides, the

Sludge of plant nr.	Lab.invest. p.e. g/kg d.s.	Selection based on MFT - % d.s. good	moderate	bad	Actual p.e. in practice at time of investigation type	g p.e./kg d.s.
aerobic. stab.						
	2,5	AD	BCF	F		
1	5	AE	F	BCD	AE	5.5
	7,5	AE	F	BCD		
	10	AE	DF	BC		
2	5	AD	BG$_2$	CFG$_1$	G$_1$	4
	7.5	A	BDG$_2$	CFG$_1$		
	5	AB	D	C		
3	7.5	AL	BD	CF		
	10	AL	C	BDF	L	10
	2,5	BH	AD	-		
4	5	ABDH	-	CF	H	4.5
	7,5	AH	D	BCF		
5	5	AK	BD	CF	K	3
	7.5	AK	D	BCF		
	2,5	ABDFH	-	C	H	3.5
6	5	ABDH	F	C		
	7.5	ADH	-	-		
	2,5	BFM	ACD	-		
8	5	ADM	BF	C	M	4,5
	7,5	A	BDFM	C		
digested	2,5	ABF	CDO	-	O	3.5
10	5	ABDFO	-	C		
	7.5	ABDF	O	C		
	2,5	AD	Z	BCF		
11	5	AD	BZ	CF	Z	4,9
	7,5	AD	BZ	CF		
12	2.5	ADL	BF	C		
	5	L	ADF	BC	L	3.6
	7,5	ADL	B	F		
14	2,2	BND	AF	C	N	3.3
	4.4	ABD	FN	C		
	6.6	AD	BFN	C		
15	2,5	BDG$_1$F	-	AC	G$_1$	2.5
	5	ABD	G$_1$F	C		

Table 3. Selection of polyelectrolytes based on laboratory research and comparison with the applied polyelectrolytes(for code see table 2.)

table gives other practical information. It appears that in many cases the dewaterability is satisfactory for more polyelectrolytes. In other words: the operator can choose from several alternatives. It will be clear that in such cases besides improvement of the dewaterability other aspects as cost price and manageableness of the product (dissolution) play a part. Because of these aspects other polyelectrolytes might be applied or selected than the one used in the sewage treatment plant.

The table also shows that several polyelectrolytes can be used over a large range of dosages.
(At the sewage treatment plant numbered 10, after the investigation a more suited polyelectrolyt (by able 3) was put into use owing to which the dry solids content of the sludge cake of the belt press has increased by 2 to or 3%).

5. MIXING OF POLYELECTROLYT AND SLUDGE

Selecting a suitable polyelectrolyt in accordance with the previous procedure does not automatically mean that the combination of p.e. and sludge is used in an optimal way. It is desirable to check regularly whether the mixture of p.e. and sludge gives the maximum output. During conditioning, insufficient mixing intensity as well as excessive mixing intensity of p.e. and sludge have an adverse effect on the dry solids content of the belt press cake. Though not all belt presses have an easily adjustable mixing device in conditioning the mixing intensity can be varied in several ways. (mixing drum - rpm / slope, static mixer, p.e. dosing-before or after sludge supply pump etc.).

5.1. Method

Mixing of p.e. and sludge in practice is studied by determining the suction time of a technically conditioned sludge preferably followed by the determination of the dry solids content with the MFT test. Repeating the measurement after extra stirring with the standard stirring method of ch. 2.1 (no addition of chemicals!) gives information on the mixing during practical conditioning. Reduction of the suction time after extra stirring on laboratory scale means that the mixing of p.e. and sludge in actual practice must be intensified. If on laboratory scale (stirring test) the suction time only increases when the stirring time increases the mixing of p.e. and sludge has been too intensive or is just optimal. Then reduce the stirring energy during conditioning in practice and once more determine the dewatering parameters.

An example is given in fig.4. It concerns technically conditioned sludge from a plant which was checked for interaction of polyelectrolyt and sludge.

Since additional stirring at laboratory scale improves both dry solids content of MFT test (higher) and suction time (measure for rate of dewatering - lower) in practice more intensive mixing is required at this dosage of polyelectrolyt.

Another method to analyse the mixing of p.e. and sludge in actual practice is to vary the mixing intensity and to measure each time the turbidity of the leaking water from the predewatering zone. When the turbidity is at its lowest the mixing of p.e. and sludge is at an optimum [lit.3].

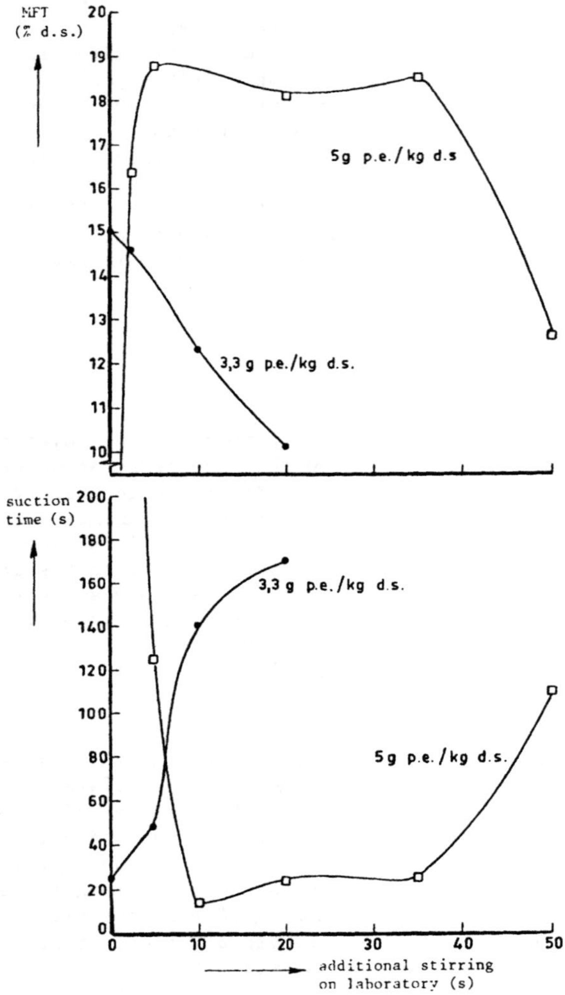

Fig.4. Check on efficient use of polyelectrolyt in practice - dewatering parameters of technically conditioned sludge

6. CONCLUSIONS

For their dewaterability on belt presses sludges can be characterized by means of laboratory conditioning using a number of dosages of Praestol 444 K for which dry solids content of the MFT test is determined. This makes it possible to compare sludges mutually, to interpret the dewatering results attained in existing plants and to predict the dewaterability of other 'unknown' sludges.

In a simple way polyelectrolytes can be selected for the dewatering of aerobically stabilized and digested sludges on belt presses. For the sludges examined, selection from a number of well-operating polyelectrolytes appears to be possible. Selection on other grounds than only the workability may be considered, such as the cost price and the manageableness of the polyelectrolyt for the polyelectrolyt solution to be used.

For conditioning sludge a variable mixing intensity is important. The optimum mixing intensity depends on the type of polyelectrolyt and the dosage used. With the MFT test and also the turbidity test it can be assessed whether the mixing intensity during conditioning in practice is perhaps insufficient or too intensive.

7. LITERATURE

1. STORA reports on sludge dewatering nrs. 1-6, address STORA: P.O.Box 414, 2280 AK Rijswijk Z.H., The Netherlands (Reports and manuals available in Dutch).

2. Heide, B.A.,R.Kampf, H.J.van Veen, M.A.Visser
Selection and use of polyelectrolytes in sludge dewatering with belt presses, H_2O, 15 (1982) 4, p.62-67 (in Dutch, English translation, publ.nr.813, available from TNO Research Institute for Environmental Hygiene, P.O.Box 214, 2600 AE Delft, The Netherlands).

2. Heide, B.A., R.Kampf, M.A.Visser
Manual for the selection and use of polyelectrolytes in dewatering sludge with belt presses, Report A 124E, Febr.1982, available from TNO, for address see 2.

4. Heide, B.A. and R.Kampf. The MFT method as characterization of the dewaterability of sludge. H_2O 11 (1978) 15, p.339-342 (in Dutch).

5. European co-operation and co-ordination in the field of scientific and technical research COST project 68-EUCO/PS/48/75, Sewage Sludge Processing (final report, November 1975).

DISCUSSION

P. Balmer

Your method of selection of polyelectrolyte is based on the MFT cake dry solids criterion. The rate of dewatering is also an important operational parameter, since it determines the output capacity of dewatering plant. How are these two parameters related?

B.A. Heide

The MFT test provides measures of both the rate of dewatering (suction time) and the cake dry solids, as may be seen in Fig. 4 of the paper. However, the MFT suction time, which can be very quickly measured, is used mainly as a preselection screening test for polyelectrolytes, while the dry solids content test, which takes longer to carry out, is used for the final selection from the short list.

H.W. Campbell

It is not clear to me how the rating system in Table 1 of the paper works. Why, for example, is sludge 8 rated at 4 and sludge 10 rated at 2? Secondly, would it not be more useful to express your results in cost terms rather than in polyelectrolyte quantity used, since there is such a wide variation in polyelectrolyte cost?

The numbers in the remarks column of Table 1 are not ratings as such, but relate to the evaluations based on the laboratory results, which are presented as footnotes to Table 1. We have not used cost as a parameter as we are mainly concerned with outlining experimental procedures for selection of polyelectrolytes. Obviously, cost should be taken into account in the final selection of polymer.

G. Mininni

Mixing time and mixing intensity are important parameters for belt-pressing of sludge. Is it possible to predict full-scale mixing requirements from laboratory tests?

It is not possible to make such a prediction. We can only verify in the laboratory whether a particular full-scale mixing system is adequate or not by checking a sample taken from the full-scale unit. In this regard, it should be remarked that it is not possible to vary the mixing input on many belt presses.

G. Hoyland

Fig. 2 of the paper shows that the dry solids concentrations obtained for aerobically digested sludges are lower than those obtained for anaerobically digested sludges. Have you plotted similar correlations for the suction time parameter?

Yes, a similar pattern is obtained for the suction time variation with polymer dose i.e. sludges which yield a high dry solids content have a low suction time. However, we prefer to use the dry solids parameter for the characterisation of sludges in respect of dewatering.

ASSESSMENT OF SLUDGE CENTRIFUGABILITY

L. SPINOSA and G. MININNI
Istituto di Ricerca sulle Acque - C.N.R. Italy

Summary

Sludge dewatering by centrifuge is notably affected by general feed sludge characteristics such as settleability, scrollability and floc strength. The measurements of settleability,alone either with traditional devices or with a stroboscopic centrifuge, did not seem to be useful in determining the conditioner dosage or in predicting full-scale equipment performance. One test was proposed by Vesilind and consists of a determination of both characteristics of settleability and scrollability. This method appeared much more efficacious, especially for laboratory prediction of centrifuge solid recovery (efficiency) and cake concentration. However, the test is not suitable for activated sludge or for those sludges not sufficiently concentrated with very poor characteristics of scrollability. Tests carried out using full-scale and pilot-scale centrifuges reveal the influence of floc strength on efficiency of centrifugation and the possibility of using this test to determine the optimal conditioner dosage for obtaining the best recovery.

1. INTRODUCTION

Centrifugation is one of the most common methods utilized for sludge dewatering. The industrial machines more widely employed in sludge treatment consist of a cylindrical conical bowl shell with an internal Archimedean screw (the scroll) which rotates at a slightly lower speed and scrapes the solids up the beach and out of the machine (fig. 1). The advantages of this technology consist in the fact that solid-liquid separation takes place in completely isolation from the outside and that the machine is relatively small,versatile and simple to operate. Two functions are required for satisfactory operation:

- production of a clean centrate
- " of a concentrated cake.

Unfortunately a parameter for assessing sludge suitability

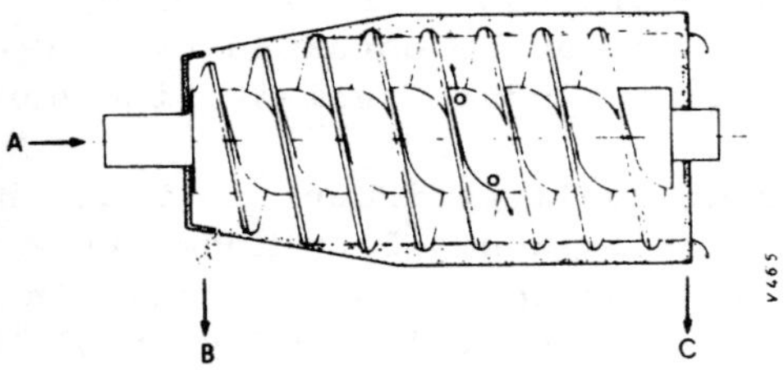

Fig. 1 - Schematic view of a bowl centrifuge
A feed sludge
B dewatered sludge
C centrate

for centrifuging has not been yet defined because it has not been possible to reproduce on a laboratory scale the conditions occurring in an industrial centrifuge. However several methods have been developed in order to:
- select the conditioner type and assess its optimal dosage;
- predict performance of industrial dewatering machine.

2. SLUDGE CHARACTERISTICS AFFECTING CENTRIFUGABILITY

Centrifugability is defined as the aptitude of the sludge to be dewatered under the action of the centrifugal force. In centrifugation the mechanism of solid/liquid separation is similar to sedimentation, but solids are subjected to forces many times greater than gravity.

Moreover in a full scale centrifugation the effects induced on the sludge by the screw movement have to be taken into account. In fact the consistency of the centrifuged sludge must be such that it can be easily conveyed by the screw.

Centrifugation performance is also affected by the aptitude of the sludge flocs to resist the mechanical stresses they are subjected to.

Centrifugability is thus seen to be affected by the following sludge characteristics:
- settleability
- scrollability
- floc strength.

A safe method to evaluate all these properties is to carry out pilot scale tests. As this procedure is expensive and not always possible, laboratory methods have been proposed for the assessment of centrifugability.

3. LABORATORY METHODS

The methods available for evaluating the above mentioned sludge characteristics are described in the following.

3.1 Evaluation of settleability

The aptitude of a sludge to settle can be evaluated by measuring the settling velocity. This can be obtained by plotting the height of the solid/liquid interface against time during settling tests in cylinder.

The main disadvantage of using small cylinders is that the measure depends on the size; in fact in a small cylinder the settling of highly concentrated suspensions is slower than in a large one because of bridging across the walls and partial support of the solids from below. To overcome these difficultie a modified test involving the settlement of sludge in a standardized stirred column has been developed (White 1975).

Bach settling tests can be conducted also at high gravitational forces using a lab stroboscopic centrifuge. This procedure has the disadvantage of depending on the force imposed (angular speed and radius); to this end Vesilind (1977) introduced a "settling coefficient" which is independent of the centrifuge speed and radius, and thus represents the sludge characteristics only.

Many other completely empirical tests are utilized from manufacturers (Vesilind 1969, Campbell et al. 1978); among them the following ones are mentioned:

- sludge spinning in a bottle centrifuge for 30 s and examination of the centrate clarity;
- sludge centrifugation in a small disc type machine.

3.2 Evaluation of scrollability

The semi-liquid material consistency can be tested with a penetrometer. The test consists in allowing a standardized needle to drop into the tested material. Similarly sludge consistency, and thus its scrolling properties, can be determined by a penetration test using the standard penetrometer shown in fig. 2 (Vesilind, 1969). This instrument consists of a polished brass rod (diameter 3 mm, weight 13.4 g) inserted in a plastic support. The test is carried out by measuring the height of cake penetrated by the rod, after centrifugation in a bench centrifug and after pouring off the centrate. The penetrability is calculated as the percentage of sludge not penetrated.

3.3 Evaluation of floc strength

This method allows the mechanical resistance of the flocs to be evaluated and thus their aptitude to resist the stresses they are subjected to in undustrial machines. The test is carried out by subjecting the sludge to stirring at 1000 rpm for different times using the standard apparatus shown in fig. 3 (Commission of the Eur. Comm. 1975) and measuring the CST of stirred sludge. The floc strength can be evaluated by plotting

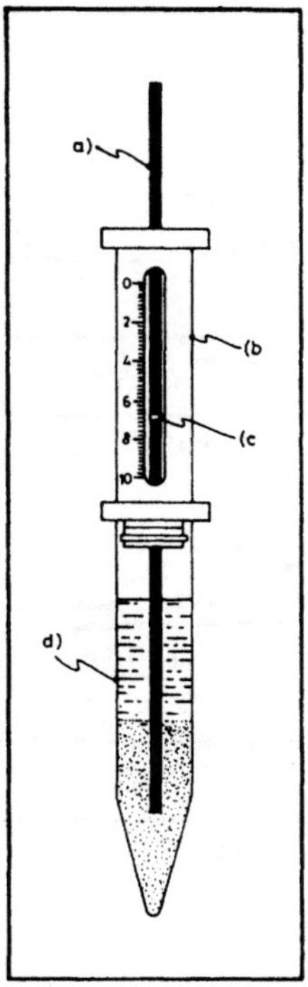

Fig. 2 - Standard penetrometer
a) brass rod
b) plastic support
c) index
d) centrate

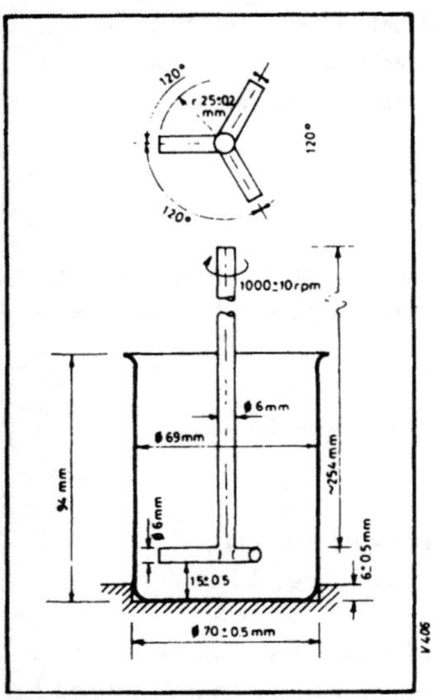

Fig. 3 - Standard stirrer

CST values vs. time of stirring. Good industrial centrifuging results can be expected when:
- the pattern of CST vs. stirring time is linear with a slight slope, between 10 and 100 s
- the CST value at 10 s stirring is around 10-12 s (with Ø 18 mm reservoir).

4. APPLICABILITY OF METHODS

Two laboratory methods are in practice available for evaluating the feasibility of industrial centrifugation and selecting the type and dosage of polymer required; the first one utilizes a lab centrifuge while the second is based on floc strength measurements. Useful information on the expected machine performance can be also obtained.

Although methods based solely on settleability measurements can be utilized, the test carried out by Mininni et al. (1982) have shown them to be comparatively useless. Experiments using slaughter-house sludges showed that settling rate in 1 l glass cylinder, at different conditioner dosages, are similar and in a few cases identical (fig. 4) in spite of the different performances obtained with the full scale equipment. Tests carried out with a low speed centrifuge led to similar conclusions (fig. 5).

4.1 Lab-scale centrifugation

This method, proposed by Vesilind (1969-1970) allows both settling and scrolling properties to be evaluated. It consists basically in centrifuging sludges by a lab centrifuge under various conditions of centrifugal force and centrifugation time. Settling properties are measured by determining the suspended solid concentration in the centrate while scrolling ones are determined by the penetration test. Correlation of results with

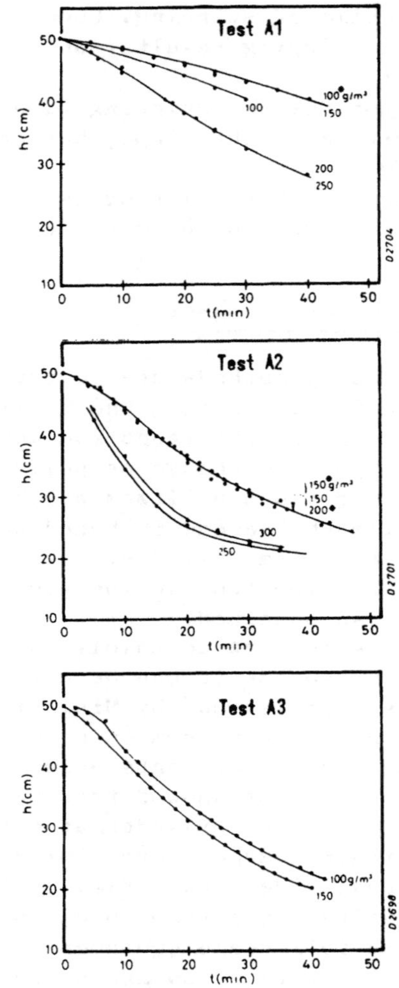

Fig. 4 - Settling curves in a 1 liter cylinder at different conditioner dosages

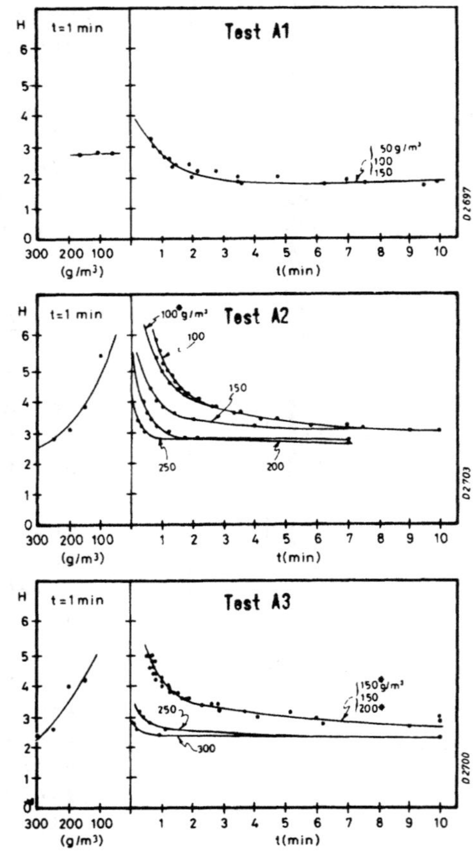

Fig. 5 - Height of solid-liquid interface in tube of stroboscopic centrifuge at different conditioner dosages

full scale installations allows the following equation to be obtained for the prediction of the solid recovery (E) achievable with full scale equipment:

$$E = \frac{C_o - C_1}{C_o} \cdot 100 \cdot \left(\frac{P}{100}\right)^{0.1} \qquad (\%)$$

where:

C_o = input sludge suspended solids concentration (mg/l)
C_1 = centrate suspended solids concentration (mg/l)
P = penetrability (%).

The first term represents the aptitude of particles to be separated from the liquid, while P is a measure of the scrollability.

Test carried out at Breda Research Institute (Aveni and Lamarca 1974; Spinosa et al. 1982) showed that, by plotting the values of $(C_o-C_1)/C_o$ and P, vs. centrifugal force at different centrifugation times on a log-prob paper, linear relationships are obtained (Fig. 6) which allow the efficiency obtainable for the same sludge under different operating conditions (Fig. 7) to be evaluated. The influence of P in evaluating centrifugability in shown in Tab. 1, which shows the results of comparative tests carried out on both laboratory and industrial scales with cellulose production sludges.

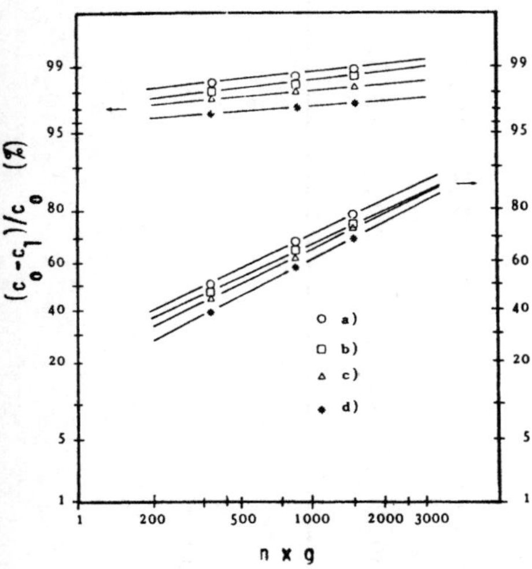

Fig. 6 - Vesilind test with cellulose production sludges
a) centrifuging time = 120 s
b) " " = 90 s
c) " " = 60 s
d) " " = 30 s

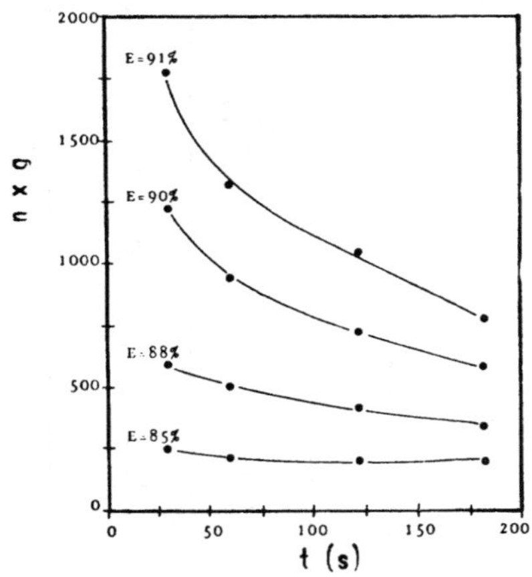

Fig. 7 - Performance evaluation of full scale equipment by Vesilind test with cellulose production sludges

The cake concentration can be predicted by laboratory tests conducted under the same conditions of centrifugal force and centrifuging time.

In practice, in many cases, this method has not provided satisfactory results with activated sludges because of the difficulty involved in evaluating the sludge consistency since the P value almost zero. This does not exclude the possibility of centrifuging the activated sludge, although cake concentrations generally do not exceed 12% (U.S. EPA, 1979) even if high solids recovery is achieved.

If the test is carried out at a single value of centrifugal force and centrifugation time, it results possible to calculate a "centrifugability index", calculated using the above relation-

Tab. 1 - Comparative centrifugation tests carried out on laboratory or full scale equipment
Assessment of applicability of Vesilind test.

Centrifug force	Centrifug time	Sludge 1			Sludge 2			Sludge 3		
		Laboratory scale		Full scale	Laboratory scale		Full scale	Laboratory scale		Full scale
(xg)	(s)	$\frac{C_o - C_1}{C_o}$ 100	E	E_{ind}	$\frac{C_o - C_1}{C_o}$ 100	E	E_{ind}	$\frac{C_o - C_1}{C_o}$ 100	E	E_{ind}
1300	10	95,0	90,0	83,8	97,2	71,9	67,0	94,0	88,0	80,2
1300	17	96,2	91,5	84,9	97,6	75,6	69,0	94,7	89,3	82,6
1250	23	96,7	92,4	87,9	97,8	77,2	70,7	95,1	90,0	85,0
1200	31	97,2	92,9	89,2	97,9	79,0	72,8	95,4	90,7	86,7

Sludge 1 ≑ Primary (SS ≑ 4,1 - 5,3%)
" 2 = Activated (SS = 2,0 - 3,7%)
" 3 = Mixed 70% 1 + 30% 2 (SS = 4,7%)

ship, which can be utilized for comparing different sludges and/or different types and dosages of conditioners.

Vesilind (1969) also reports a similar method consisting in sludge spinning in a bottle centrifuge for 22 s, qualitative visualization of the centrate and investigation of the cake consistency by prodding it with a glass rod.

A similar method has been proposed by Campbell et al. (1978). It consists in centrifuging 50 ml samples in a bench centrifuge at about 1000 g for 120 s; a portion of the sludge cake is scrolled out of the tube using an anger drill bit while the tube is held at a 45° angle. Solids analyses are carried out on the centrate and on the scrolled portion: another value of cake concentration is obtained on the basis of the cake volume before scrolling.

4.2 Floc strength

Tests were carried out at the Water Research Institute to ascertain whether the floc strength method can be useful in assessing the optimal conditioner dosage and determining the influence of floc strength on pilot or full scale equipment performance. The sludge was conditioned in the laboratory by adding under standard conditions (10 s for polyelectrolyte addition while stirring at 500 rpm and 20 s of additional mixing for flocculation) a volume of conditioning solution always amounting to 10% of that of the sludge. Sludge floc strength was then evaluated at different chemical dosages. Parallel tests with industrial equipment were conducted at the same dosages of laboratory experiments to verify the applicability of this method. In one case (sludge A) a large centrifuge was used (bowl diameter 353 mm) and in another (sludge B) a small one (152 mm). Results are summarized in tab. 2.

For sludge A solids recoveries higher than 90% were obtained at the dosages assessed in laboratory (Mininni et al. 1982). Floc strength curves are shown in Fig. 8. Cake concentrations were good and did not seem to be significantly influenced by conditioning.

Cake concentration was found to be significantly influenced by the sludge residence time on the beach and by the hydraulic or solids residence times, while the initial characteristics of the sludge has little effect.

Sludge CST was found to have a significant effect on solids recovery, while liquid ring height is the most important operational machine variable.

Sludge B proved to be unsuitable for dewatering by centrifuge with the polyelectrolyte used, because the laboratory test results did not meet the requirements mentioned. In fact a linear pattern of CST vs. stirring time corresponded to very low CST values, i.e. the same as water alone and was therefore

Tab. 2 - Comparative centrifugation tests carried out on full or pilot scaleequipment. Assessment of applicability of floc-strength test.

Sludge	Test	Lab. opt. dosage (g/m^3)	Actual dosage (g/m^3)	Input sludge flowrate (l/h)	Input sludge conc. (%)	Liquid ring height (mm)	Different. speed (rpm)	Final conc. (%)	Solids recovery (%)
A (Mixed sludge from slaughter house wastewater treatment plant)	A 1	120-170	0	9000	1.3	73.0	22	20.2	66.6
			50	9000	1.4	73.0	22	18.3	82.7
			100	9000	1.3	73.0	22	18.7	95.8
			150	9000	1.3	73.0	22	18.5	97.7
	A 2	200-250	0	9000	2.5	66.5	22	23.2	71.2
			100	9000	2.3	66.5	22	22.8	84.1
			150	9000	2.6	66.5	22	21.0	89.4
			200	9000	2.1	66.5	22	20.8	93.8
			250	9000	2.1	66.5	22	21.4	97.8
	A 3	250-300	150	9000	2.0	59.0	22	20.8	80.9
			250	9000	1.8	59.0	22	20.7	88.5
			300	9000	1.8	59.0	22	22.2	98.4
B (Mixed sludge from municipal plant)	B 1	90 (?)	0	492	1.1	11.1	30.9	15.3	31.4
			0	492	1.1	11.1	6.4	25.2	36.3
			47	492	1.1	11.1	30.9	8.2	58.3
			47	492	1.1	11.1	6.4	3.2	99.3
			94	492	1.1	11.1	30.9	5.9	89.2
			94	492	1.1	11.1	6.4	2.0	98.3
			195	492	1.1	11.1	30.9	6.0	98.5
			195	492	1.1	11.1	6.4	2.5	97.9

(cont. Tab. 2)

		0	241	1.1	4.8	30.9	23.3	25.1
		0	241	1.1	4.8	8.7	23.1	29.3
		48	241	1.1	4.8	30.9	14.8	29.6
B 2	90 (?)	48	241	1.1	4.8	8.7	12.8	38.3
		95	241	1.1	4.8	30.9	12.5	39.0
		95	241	1.1	4.8	8.7	9.7	53.7
		191	241	1.1	4.8	30.9	8.5	24.9
		191	241	1.1	4.8	8.7	12.1	99.6

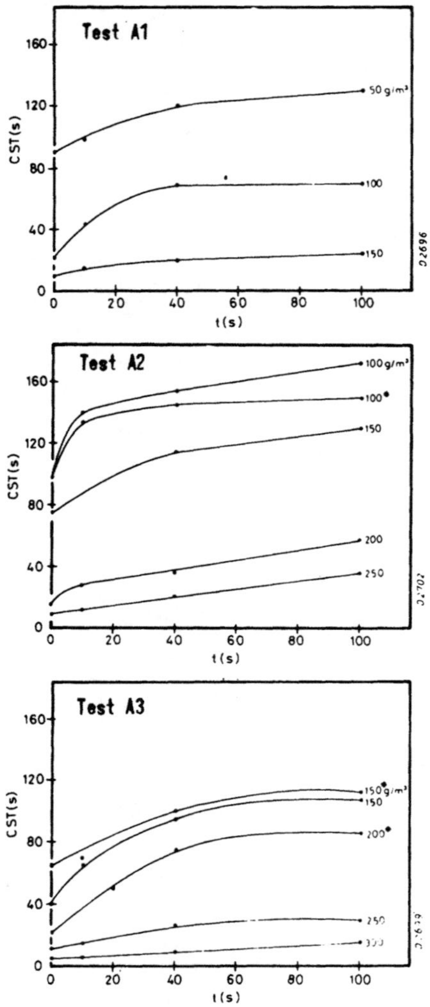

Fig. 8 - Floc strength at different conditioner dosages

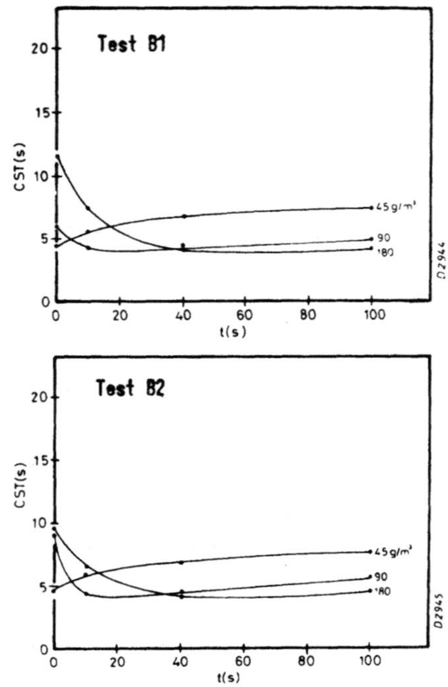

Fig. 9 - Floc strength at different conditioner dosages

not significant (fig. 9). Experiments with the industrial centrifuge also indicated very poor results in that very good solids recovery was obtained only when coupled to low cake concentrations, which were found to be considerably reduced by conditioning.

In Tab. 3 the optimal dosages determined in laboratory and the actual ones adopted in centrifuging different kinds of sludges are compared (Mininni et al. 1980; Di Pinto et al. 1981). Actual dosages are found to be slightly higher in some cases, probably as a result of less efficient sludge/conditioner mixing conditions realized in full scale tests.

A similar principle is applied in the method based on rhe-

Tab. 3 - Cecking tests for the optimal conditioner dosage.

Sludge	Test	Lab. opt. dosage (g/m^3)	Actual dosage (g/m^3)	Solids recovery (%)
	C1 (Municipal aer. stab.)	100-150 175	110 128-192	99 99
	C2 (as C1 with simult. dephosph. by $Al_2(SO_4)_3$	100-150 150-175	150-160 173-184	99 98
	C3 (as C1 with simult. dephosph. by $FeSO_4$)	150-200 150-200	208-231 159-173	91-99 98
	D1 (Munic. oxid. with pure-oxyg.)	100 100	100 115	90-99 85-88
	D2 (as D1)	250 250	255 245	87-95 90
	D3 (Munic. oxid. with air)	100-125 150	100 147-155	92-99 93-99

ological measurements. The sludge is stirred using a properly modified rotational rheometer (Colin et al. 1976) and the torque variation plotted vs. time. From the interpretation of the above curves indications similar to the previous ones are obtained. However the test requires the use of a device which laboratories are not always equipped with.

5. CONCLUSION

Centrifugability characterization of sludges is very difficult as it is practically impossible to reproduce in laboratory the conditions actually existing in an industrial centrifuge.

In industrial centrifugation three properties have to be taken into account:

- settling,
- scrolling,
- floc strength.

Two methods can be used to evaluate these properties and select the type and determine the dosage of polyelectrolyte.

In the first method settling properties are measured by determining the suspended solid concentration in the centrate after centrifuging by bench centrifuge while scrolling properties are determined by a penetration test on the cake collected in the tube. This method allows optimal conditioner dosage to be determined and the cake concentration and the solids recovery to be predicted. However in many cases it is hard to apply to activated sludges because of their poor consistency and, therefore, of difficulty in evaluating their penetrability.

The other method is based on floc strength measurements. This test seems to be a simple and reliable method for determining the optimal conditioner dosage at which a good industrial equipment performance can be expected. Moreover floc strength appears to influence solid recovery: good correlation has been obtained for slaughter house sludge.

7. REFERENCES

Aveni A., Lamarca V. (1974): "Cooperative program on the characterization of the sludges: centrifugability activity of the Istituto di Ricerche Breda" EEC COST 68 bis, Working Party 1, Dubendorf, 4-5 April 1974.

Campbell H.W., Rush R.J., Tew R. (1978): "Sludge dewatering design manual" Training of Technology Transfer Division Environmental Protection Service Fisheries and Environment - Canada

Colin F., Cornier S.C., Daniel J.L., Jacquart J.C., Lefort D., Mathiau R., Braunstein .P. (1976): "Characterization des boues residuaires" Techniques et Sciences Municipales no. 1-76 Supplement.

Commission of the European Communities (1975): "Sewage sludge processing" Cost Project 68 - Final Report of the Management Committee - November 1975.

Di Pinto A.C., Mappelli P., Mininni G., Spinosa L., Tandoi V. (1981): "Analisi delle prestazioni di un impianto a fanghi attivati con funzionamento ad aria o ad ossigeno" Istituto di Ricerca sulle Acque - C.N.R. Rapporto Interno R/80.

Mininni G., Santori M., Spinosa L. (1980): "Problemi di condizionamento e disidratazione dei fanghi di un impianto convenzionale di depurazione con rimozione chimica del fosforo" Istituto di Ricerca sulle Acque - C.N.R. Rapporto Interno R/78.

Mininni G., Spinosa L., Scalici E. (1982): "A study on sludge centrifugability" EEC Cost 68 ter, WP 1 Meeting Zurich 13 May 1982.

Spinosa L., Aveni A., Lamarca V. (1982): "Valutazione e applicazione dei parametri di caratterizzazione tecnologica dei fanghi" Seminario sulla Disidratazione Meccanica dei Fanghi, IRSA-C.N.R. Bari 24-25 novembre 1982.

U.S. Environmental Protection Agency (1979): "Process design manual for sludge treatment and disposal" EPA 625/1-79-011 September 1979.

Vesilind P.A. (1969): "Estimation of sludge centrifuge performance from laboratory experiments" NIVA Project no. T3/69.

Vesilind P.A. (1970): "Estimation of sludge centrifuge performance" J. Sanit. Engng. Div.-ASCE, 96 (SA3) 805.

Vesilind P.A. (1977): "Characterizing sludge for centrifugal dewatering" Filtr. & Separn., 14, 115.

White M.J.D. (1975): "Instruction manual for WRC settling apparatus for activated sludge" WRC Technical Memorandum TM 103.

DISCUSSION

B.A. Heide

We have found that a high non-volatile solids fraction appears to aid sludge dewatering by centrifuge. Did you also observe this behaviour with the three sludge types for which experimental results are presented in Table 3 of your paper?

L. Spinosa

We did not examine these sludges from this viewpoint. However, I can say that sludge C1 had better dewatering characteristics than sludges C2 and C3, which contained aluminium and iron additions, respectively. This difference was, however, due in part to loading differences at the plants where the sludges were produced. I cannot therefore confirm your observation on the influence of the non-volatile solids on centrifuge dewatering.

P.H. Brunner

The equation for solids recovery presented in your paper includes the parameter sludge penetrability. Is there a theoretical reason for this or on what basis has it been included?

This equation is an empirical/statistical equation which has been established by statistical analysis of experimental data. The influence of sludge penetrability of solids recovery is illustrated in Fig. 6 in the paper.

B. Paulsrud

You mentioned that the penetrometer test is not appropriate for predicting solids recovery for activated sludges. In our experience the penetrometer test has not been found to give reliable prediction of solids recovery, even under conditions where the solids recovery exceeded 90%. Could you comment on this in the light of your experience.

Good results have been obtained with chemical sludges, as the results in Figs. 6 and 7 and Table 1 show. With activated sludges, good results can only be obtained if the initial solids concentration is high, say 4 to 5% solids, but the method is not suitable for thin activated sludges having a concentration in the range 1 to 2% solids.

A.M. Bruce

Could you express your opinion as to whether or not there is need for further test development on sludge centrifugation.

I think the tests now available are adequate but they need to be further verified by more widespread use. We intent to terminate our research in this area in the coming year.

ASSESSMENT OF SLUDGE CONDITIONABILITY USING RHEOLOGICAL PROPERTIES

H.W. CAMPBELL and P.J. CRESCUOLO
Environment Canada
Environmental Protection Service
Wastewater Technology Centre

Summary

The second phase of an ongoing research program at the Wastewater Technology Centre was conducted to evaluate the applicability of using rheological measurements as control parameters for sludge conditioning, and to develop a methodology for the on-line measurement of sludge rheology from a continuous-flow, pilot-scale chemical conditioning system.

A rotational co-axial viscometer was interfaced with a computer to facilitate data logging, data analysis and ultimately computerized control of the viscometer.

The relationships between rheological properties and polymer addition were examined for three anaerobically digested sludges using batch, bench-scale experiments. The sensitivity of rheology to changes in solids concentration and sludge dewaterability, was established.

A continuous-flow, pilot-scale sludge conditioning system was designed and fabricated. Experiments using this system confirmed that the relationships observed in the batch, bench-scale tests were also valid for the pilot-scale continuous system.

BACKGROUND

The natural growth of urban centres results in a need for increased sewage treatment plant capacity, which invariably results in the production of more sludge. Concurrent with this, is a trend towards upgrading existing plants by the addition of tertiary treatment processes such as phosphorus removal, which also increases sludge production. In light of these developments, sludge volume reduction (dewatering) is becoming increasingly important as an integral component of sludge handling and disposal systems.

With few exceptions, a sewage sludge dewatering system must be preceeded by some form of conditioning in order to flocculate the sludge and enhance the ease with which water may be removed. Chemical conditioning of sludge represents a significant operation and maintenance cost, and directly affects the quality of the sludge cake produced by the dewatering device. It is therefore important, from both a cost and a performance consideration, to optimize the addition of chemicals for sludge conditioning. A variety of tests, including capillary suction time (CST), specific resistance, solids concentration, etc., are available to measure the effectiveness of a particular conditioning option. Unfortunately, most of these test procedures have not been adapted to continuous on-line measurement. Rheology appears to have the potential to adequately describe the effects of chemical conditioning, and what is possibly more important, equipment is available for the on-line measurement of rheological properties.

Rheology can be defined as the study of the properties and behaviour of matter in the fluid state. Most single-phase fluids exhibit Newtonian behaviour in that the rate of viscous flow is proportional to the shear stress. The addition of solid particles to the fluid interferes with the free flow of the dispersion medium to a degree that is dependent on the rate of shear. Dispersed systems such as these can exhibit a variety of rheological behaviours including plastic, pseudoplastic and altered Newtonian. Most sewage sludges have been interpreted as exhibiting either plastic or pseudoplastic flow (Dick, 1979). They may or may not possess an initial characteristic yield stress depending on the shape of the flow curve and the definition of yield stress.

The rheology of sewage sludge is further complicated by the fact that most sludges are also thixotropic, meaning they possess an internal structure which breaks down as a function of time and shear rate. A flow curve or rheogram (Figure 1) of a typical sludge, showing the curves produced by a rotating viscometer during the increasing and decreasing rate of shear cycles, indicates that the rheology of the sludge has been altered during the initial (increasing) phase of the test. The displacement of the two curves is referred to as a hysteresis loop and is a measure of the degree of thixotropy exhibited by the sludge.

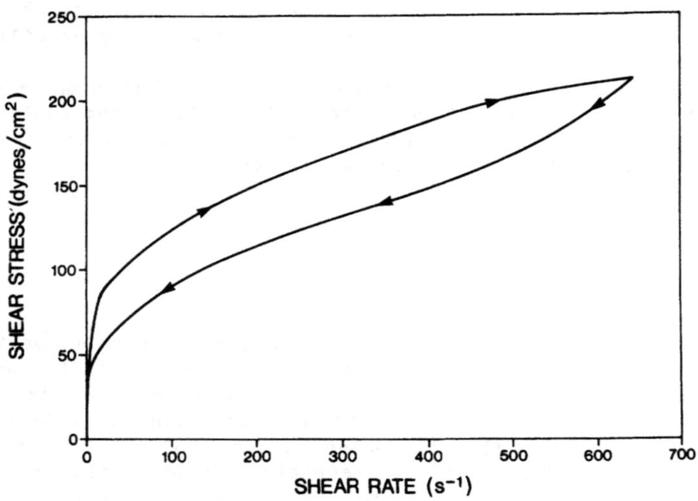

Figure 1. Typical rheogram illustrating thixotropy.

SUMMARY OF PREVIOUS WORK

The current rheological program at the Wastewater Technology Centre is based on conclusions drawn from rheograms (Figure 2) developed during previous sludge characterization studies (Campbell et al, 1978). The basic observations were two-fold. The initial shear stress or yield stress, increased as the polymer dosage increased. The second observation was that the polymer dosage (6 kg/t) at which the characteristic shape of the curve changed (i.e., an initial peak became evident) compared favourably with the polymer dosage required to produce efficient performance in concurrent pilot-scale dewatering trials. Details of the first portion of the study (i.e., to evaluate the potential use of rheological measurements) have been reported in the literature (Campbell and Crescuolo, 1982) but will be summarized here briefly.

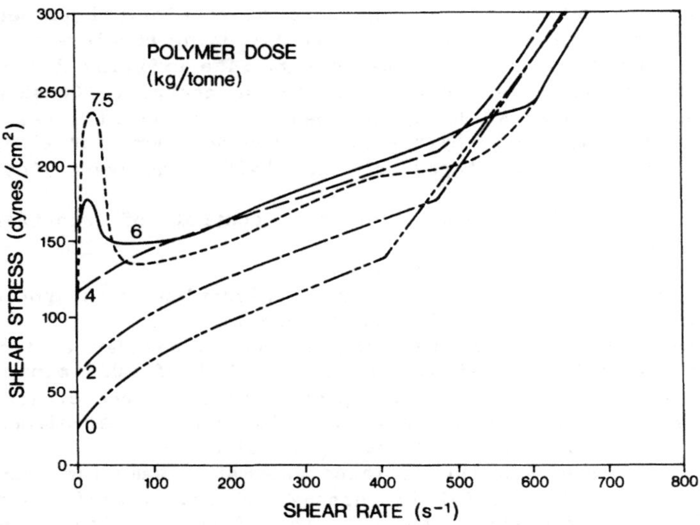

Figure 2. Rheograms of polymer conditioned sludge.

The development of a standardized, laboratory-scale methodology for measuring the rheological properties of chemically conditioned anaerobically digested sludges was the first priority. The method selected for adding chemicals (polymers) to the sludge, consisted of adding the required volume of polymer over a 10-second period, while mixing the sludge at 1000 rpm by a standard stirrer developed by the Water Research Centre (Cost-Project 68, 1975). A Haake Rotovisco RV3 rotational viscometer with an MVII sensor system (gap width = 2.5 mm) was used for all rheological measurements. The acceleration rate was standardized at 75 rpm/min and the damping option was not utilized.

The concepts of yield stress and apparent viscosity were discussed and it was concluded that these concepts were not particularly appropriate to chemically conditioned sludges which exhibit a peak in the rheogram. It was suggested that a more useful parameter might be defined as the "instantaneous viscosity". This parameter would be the derivative of the flow curve and would have the units of viscosity. The peak in the rheogram could then be identified as the point at which the derivative was zero.

The effect of shear forces on the particle size distribution of conditioned sludges was also studied. It was determined that the breakdown of flocculated sludge particles was a function of the level of conditioning, shear rate and shear time. Increasing any of these variables tended to increase the degree of particle breakdown.

This paper describes the current status of the ongoing program. The major objectives of this phase of the study were as follows:

1. To evaluate the applicability of using rheological measurements as control parameters for sludge conditioning;
2. To develop a methodology for the on-line measurement of the rheological properties of sludge from a continuous-flow, pilot-scale chemical conditioning system.

RESULTS AND DISCUSSION

Equipment Upgrading and Computerization

Throughout the study, all of the data on rheological properties had been generated from the viewpoint of eventually using rheology as the basis for automatic control of sludge conditioning. The existing RV3 rotational viscometer was not suitable for on-line process control. Consequently an upgraded model, the Haake RV100 was purchased. Side-by-side testing indicated that the data generated by the two viscometers were for all practical purposes identical. Advantages of the RV100, however, include the following:

- Capability of external computer control of start/stop/pause functions;
- Greater flexibility for data display;
- Continuous rather than step-wise acceleration of the rotor;
- Greater sensitivity at low shear rates.

In conjunction with the upgrading of the viscometer equipment, the viscometer was also interfaced with a Hewlett Packard 1000 mini-computer. This allowed control of the start/stop/pause functions from either a remote terminal or from a real-time program. This is an essential element of any control strategy which involves the use of a viscometer.

During a run, the computer collects data at a sampling frequency of 10 times/second for the initial 18 seconds and four times/second for the remainder of the run and stores the data in a file. Manipulation of this data file makes it possible to regenerate the rheogram using expanded axes and also to develop the derivative of the flow curve (i.e., instantaneous viscosity) by simply calculating the slope between each consecutive pair of data points. Figure 3 shows the rheogram produced by the viscometer x-y plotter and the rheogram generated by re-plotting the data stored in the computer file. The two curves have been offset vertically for comparison purposes. When superimposed on each other they are indistinguishable.

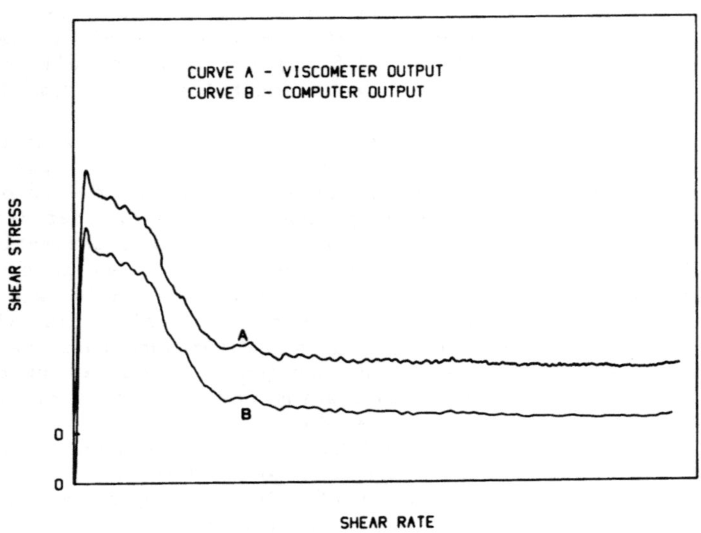

Figure 3. Comparison of viscometer and computer output.

The level of noise present in the curves is typical of the results generally obtained. In cases where the noise was significantly greater, the run was repeated. The rheograms in the balance of this report have been drawn as smooth curves to facilitate interpretation when there are several curves on one figure.

Figure 4 shows the derivative of the flow curve for the set of data shown in Figure 3. The abrupt change in the derivative from positive to negative at approximately 3 s^{-1} indicates the presence of a peak in the rheogram (Figure 3). A computer search, within prescribed boundary conditions, of the derivative data file would identify that a condition of zero slope did exist, and therefore a necessary requirement for optimal conditioning, i.e., a peak in the rheogram must have been satisfied.

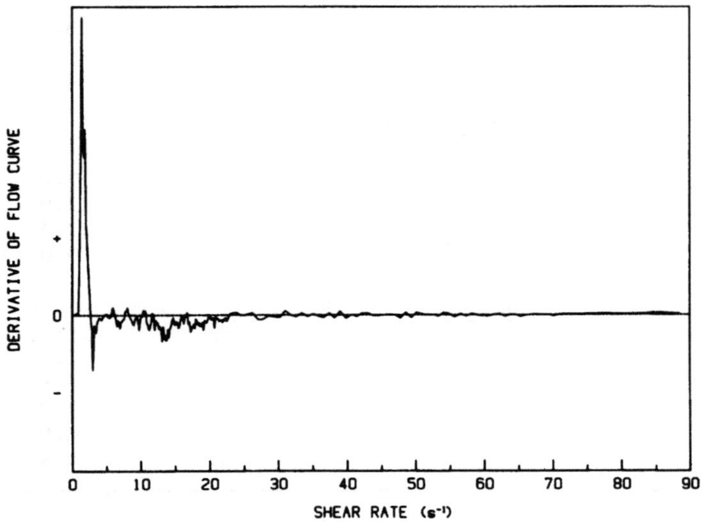

Figure 4. Derivative of sludge rheogram.

Rheological Characteristics of Polymer Conditioned Sludges

The rheological data base generated at the Wastewater Technology Centre had shown that all municipal sludges would develop a peak in the rheogram at some specific polymer dosage. However, the test conditions were not always consistent and it was difficult to make comparisons between sludges. In this phase of the study, it was decided to evaluate the effect of polymer addition on the rheology of three anaerobically digested sludges using the standardized methodology outlined by Campbell and Crescuolo (1982). A high molecular weight, cationic polymer (Praestol 429K) was used for all tests. Table 1 contains characterization data for the sludges as a function of polymer dosage. The CST's were conducted using the 10 mm reservoir. The centrifuge cake solids were determined by spinning samples of sludge in a laboratory centrifuge for two minutes at approximately 1000 G's, decanting the supernatant and analyzing the remaining cake for total solids content.

The rheograms for the three sludges are shown in Figures 5, 6 and 7. Since all of the sludges were at comparable solids concentrations (4.05 to 4.79%) differences in the rheograms can be attributed to inherent sludge characteristics. The major difference is that sludges T and H require polymer additions of only 2 kg/t to produce a peak, whereas sludge G requires 4 kg/t. At the same time, the shear stress at any shear rate for unconditioned sludge G is measurably higher than for either of the other unconditioned sludges, even though sludge G has the lowest solids concentration.

TABLE 1. Sludge Characterization Data

Polymer Dosage (kg/t)	CST (s)	Centrifuge Cake Solids (%)
	SLUDGE T, 4.79% TS	
0	2000	6.1
1	97	8.9
2	16	10.4
2.5	17	12.3
	SLUDGE H, 4.10 % TS	
0	544	7.0
1	27	11.0
2	14	12.3
3	14	12.0
	SLUDGE G, 4.05% TS	
0	1373	5.0
1	735	6.6
2	537	7.8
3	153	6.9
4	20	9.5

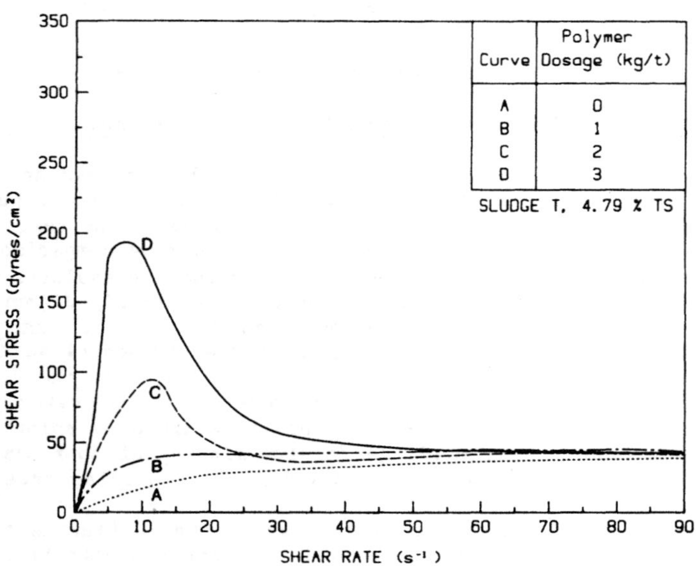

Figure 5. Effect of polymer dosage on rheology of sludge T (bench-scale).

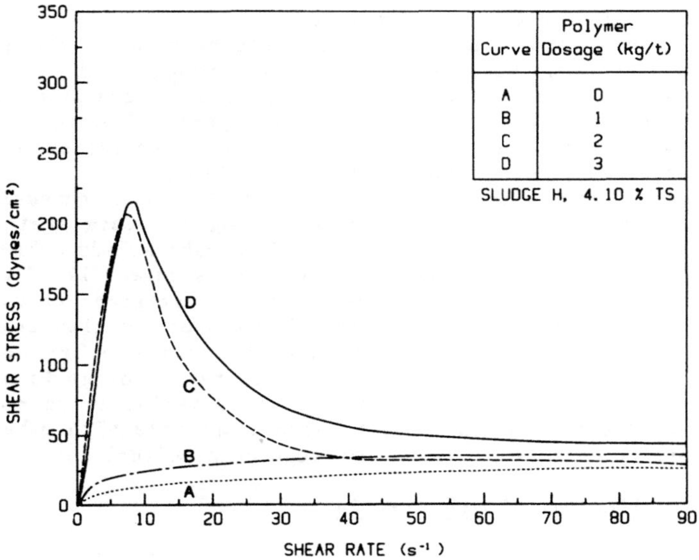

Figure 6. Effect of polymer dosage on rheology of sludge H (bench-scale).

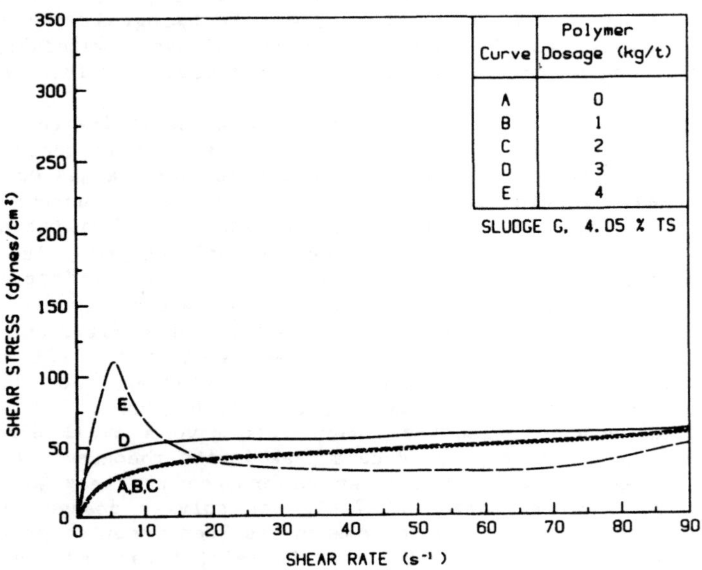

Figure 7. Effect of polymer dosage on rheology of sludge G (bench-scale).

The difference in response to polymer dosage is also illustrated by the CST data in Table 1. The unconditioned CST's for sludges T and H are higher (2000 s) and lower (544 s) than the CST for sludge G (1373 s). Sludges T and H show a dramatic decrease in CST with the addition of 1 kg/t of polymer, while sludge G shows a much more gradual reduction (almost linear) in CST with increasing polymer addition.

Based on past experience and visual observation of the conditioned sludge, it was assumed that a CST value in the range of 10 to 20 seconds, represented a super flocculated sludge. This is a condition which must normally be satisfied for dewatering by a belt press. Reference to Table 1 indicates that on this basis (CST $\leq$20 s) super-flocculation of all three sludges was achieved, but at different polymer dosages.

The trend of increasing centrifuge cake solids with increasing polymer dosage is, in general, consistent for all sludges. Once again, however, sludge G is unique in the fact that the maximum cake solids achieved (9.5%) were significantly less than those for the other sludges (12.3%).

It is apparent that sludge G must have some inherent characteristics which result in a different response to polymer addition than that observed for sludges T and H. Although the specific nature of the differences between sludges is not known, it would appear that rheology is an adequate means of describing the effect of sludge characteristics on conditioning requirements. The rheological data is also compatible with other measures of dewaterability such as CST and centrifuge cake solids.

Sensitivity of Rheological Measurements

The need for automatic process control of sludge conditioning processes arises from the fact that changes can occur in the sludge feed which require a corresponding adjustment to the polymer addition rate, if the two extremes of inadequate conditioning and overdosing are to be avoided. On a micro-scale, there are a wide variety of sludge parameters that may vary over time, but on a macro-scale there are only two gross parameters that may change, namely, solids concentration and sludge dewaterability. Any measurement intended for use in process control must therefore be sensitive to changes in these gross parameters.

All three sludges (T,H and G) were tested to determine the sensitivity of rheological measurements to changes in either solids concentration or sludge dewaterability. Initial solids levels were adjusted by either diluting or gravity thickening the sludges to achieve a number of concentrations. The dewaterability of the sludges was altered by the addition of small amounts (0.15 to 0.30% by volume) of deflocculant, in this case, household detergent. The amount of deflocculant was critical, in that excessive amounts completely destroyed the floc and no amount of polymer could restore it. The general trends observed with different levels of solids concentration and dewaterability were similar for all sludges but the magnitude of the effect varied with the sludge. For illustration purposes, only the rheograms for one sludge will be shown.

The solids concentration was observed to have a small effect on the amount of polymer required to produce a peak in the rheogram. For example, sludge T required 2.0 kg/t of polymer to produce the peak at 3.3% total solids, but at a concentration of 7.6%, the polymer dosage required was between 2.0 and 3.0 kg/t. This phenomenon has been reported previously for a physical-chemical sludge (Campbell and LeClair, 1975) and was attributed to the difficulty of dispersing the polymer uniformly throughout the sludge at higher concentrations.

The effect of solids concentration on the rheology of an unconditioned sludge (H) is shown in Figure 8. The figure illustrates the sensitivity of rheological measurements to solids concentrations and also addresses concerns which were raised when the paper describing the first phase of the study was presented (Campbell and Crescuolo, 1982). It was suggested that the occurrence of a peak in the rheogram could be simply a function of solids concentration, i.e., a peak could be generated for unconditioned sludge if the solids concentration was sufficiently high. The data shown in Figure 8 does not support this theory. Solids concentrations as high as 14% were tested and no indications of a peak were observed under the test conditions used.

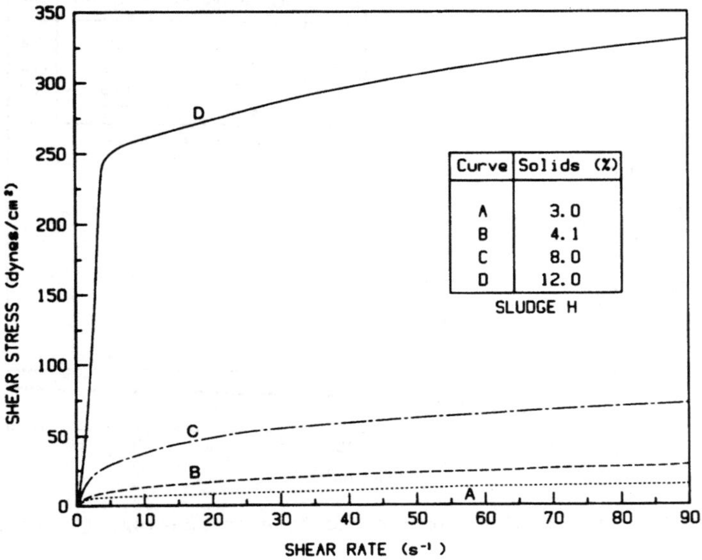

Figure 8. Effect of solids concentration on rheology (bench-scale).

The appearance of a peak, similar to those shown in Figures 5, 6 and 7, has been reported by Rimkus and Heil (1975) for concentrated digested sludge. The sludge was a 13% slurry from a sludge lagoon, and the maximum shear stress recorded at the peak was approximately 650 $dynes/cm^2$. An explanation for the apparent discrepancy between the lagoon sludge and the data shown in Figure 8 may lie in the fact that the lagoon sludge would have been exposed to freeze-thaw cycles. During a study on the applicability of natural freeze-thaw as a conditioning process, it was found that digested sludge, after conditioning by freeze-thaw, exhibited a peak in the rheogram similar to that occurring as a result of chemical conditioning (Wastewater Technology Centre, 1975). This is thought to be due to the change in the physical structure of the sludge after freeze-thaw, i.e., the presence of large aggregate particles similar to those produced by chemical conditioning.

The effect of detergent (deflocculant) addition on the rheology of sludge H is shown in Figure 9. Curve A is the rheogram for optimal conditioning of the sludge (polymer dosage of 2 kg/t and a CST of 14 seconds). Curve B shows the effect of adding detergent at a rate of 0.3% by volume. At the same polymer dosage (2 kg/t) the CST increased from 14 to 110 seconds, and the rheogram took on the appearance of that for an unconditioned sample. For curve C, the polymer addition was increased to 3 kg/t and while the CST dropped to 23 seconds, there was no significant change in the rheology. At a polymer dosage of 4 kg/t (curve D) the CST dropped to a value comparable with that for curve A, and the peak, signifying optimal conditioning, was again evident. If it is assumed that the addition of detergent is a valid means of altering the dewaterability of a sludge, then it is clear that rheological measurements can identify not only when this situation occurs, but also when sufficient extra polymer has been added to rectify the problem.

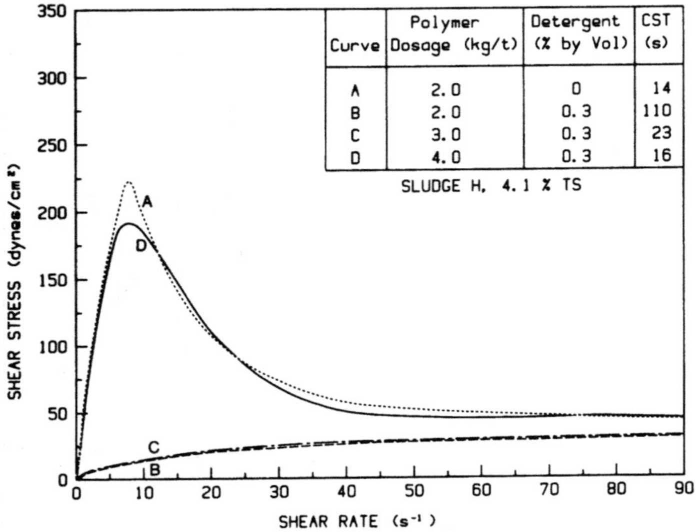

Figure 9. Effect of detergent addition on rheology (bench-scale).

Development of a Pilot-Scale Conditioning System

The development of the pilot-scale system consisted of two components; the identification of a viscosity sensor that would function on-line, and the fabrication of equipment that would condition sludge on a continuous rather than a batch basis.

The sensor system selected was comparable to that used in the batch experiments, in that it was an integral cup and rotor system. The differences were that the gap was marginally wider (2.6 mm as compared to 2.5 mm), the geometry was slightly different (length and diameter), and the cup had four slots in the wall and several holes in the bottom. The batch sensor was operated by putting the sample in the cup and then inserting the rotor. In the pilot-scale sensor system the cup and rotor are assembled as a unit attached to the motor drive. The cup/rotor portion is immersed in an open container of sludge where the sludge flows through the slots and holes in the cup, and fills up the annular space between the cup and the rotor.

Since the geometries of the batch and continuous sensors were different, it was necessary to conduct comparison runs to ensure that equivalent results could be produced by the two sensor systems. In order to eliminate any non-homogeneity between sludge samples which might be introduced through the sampling procedure, initial comparison runs were done using ketchup. Ketchup is a thixotropic material which exhibits rheological characteristics very similar to those of unconditioned digested sludge, but is more homogeneous. Figure 10 shows a comparison of the rheograms generated for ketchup using the two sensor systems. For all practical purposes the curves may be considered to be identical. Additional runs using digested sludge indicated the same degree of comparison between the sensors. It should be noted when assessing data from

Figures 10, 12, 13 and 14 that the scale of the y-axis is not consistent. Where possible, the y-axis scale was expanded in order to maximize the difference between curves.

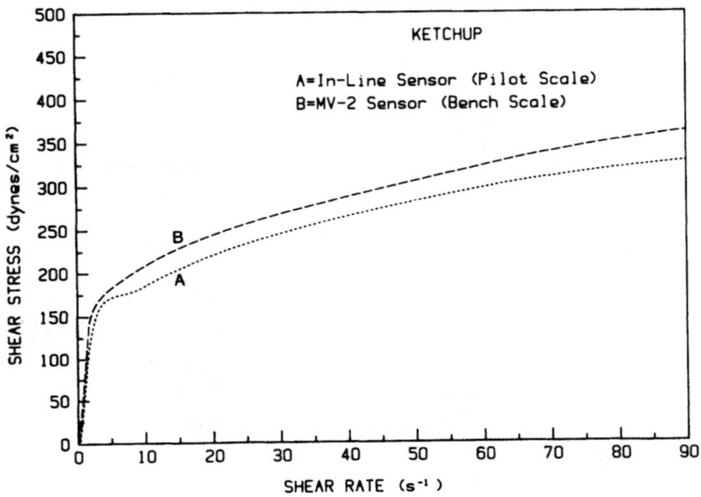

Figure 10. Comparison of outputs from bench and pilot-scale sensors.

A schematic drawing of the pilot-scale conditioning set-up is shown in Figure 11. The sludge storage tank was capable of holding approximately 1200 L of sludge and was equipped with a variable speed mixer to prevent sedimentation. The sludge pump was a variable speed positive displacement Moyno pump with a range of 3 to 22 L/min. Solids concentration was varied by the addition of dilution water to the suction side of the sludge pump. Polymer was injected into the sludge line just prior to the in-line mixer. The in-line mixer was housed in a clear PVC pipe section (47 mm internal diameter, 470 mm long) and consisted of a series of six stainless steel, helical elements welded together. Each element was approximately 76 mm long and rotated the flow through 180°, with adjacent elements being offset by 90°. The sludge flowed into the flocculator through the bottom of the tank and drained through an overflow pipe. A mixer was provided to ensure good mixing of the conditioned sludge. The viscometer sensor system was attached to a bracket such that it could be lowered into the conditioned sludge in the flocculator.

Preliminary test results indicated that the effects observed at a flow rate of 5 L/min were similar to but more pronounced than those at 20 L/min. In order to minimize the volume of sludge required, all subsequent tests were conducted at 5 L/min. The concentration of the polymer solution was adjusted such that the polymer dosing rate never exceeded 10% of the sludge flow rate. At a sludge flow rate of 5 L/min and a polymer flow rate of 0.5 L/min the retention time in the in-line mixer was approximately nine seconds.

During the development of this phase of the program, it was realized that the method employed for mixing polymer with sludge has a significant impact on the performance of the conditioning process. The pilot-scale conditioning system used in the study does not necessarily represent an optimum method of conditioning. The major criterion of the system was to produce a well-flocculated sludge at reasonable polymer dosages. The equipment, as designed, satisfies this criterion.

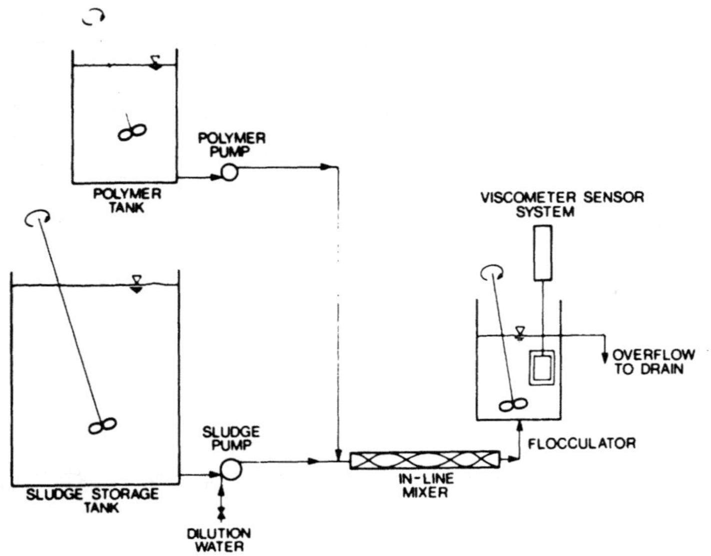

Figure 11. Pilot-scale, continuous-flow, sludge conditioning system.

The initial operating procedure consisted of setting the sludge, polymer and dilution water flow rates at the appropriate levels, waiting until an equilibrium condition had been achieved (approximately five flocculator tank retention times), and measuring the rheology of the conditioned sludge. It became immediately apparent that there were two problems associated with this methodology. The first was that the mixer in the flocculator tank created too much turbulence to allow the generation of a valid rheogram. This was solved by simply switching off the mixer during the viscometer run.

The second problem was that when the sludge was highly flocculated, the flocs tended to flow around the outside of the viscometer rather than through the slots into the sensor system. An inverted plastic container was placed over the cup/rotor assembly such that the bottom was completely open and there was an annular ring around the sensor approximately 3 cm wide. The holes in the bottom of the sensor cup were also enlarged. As the sensor system was lowered into the sludge, the outer ring filled immediately with sludge, and the sludge then appeared to flow into the sensor without any further difficulty. The rheograms generated with the system were quite similar to those generated with the batch sensor system.

The experiments conducted on the pilot-scale system were designed to determine the effect of polymer dosage, solids concentration and detergent addition on sludge rheology, and to compare these results with those obtained on the batch system. Sludge H was used for all of the experiments.

Figure 12 shows the effect of polymer dosage on rheology when the sludge is conditioned on a continuous basis. These results can be compared with Figure 6 which shows the same experiment done on a batch basis. The trends in the two experiments are identical. The peak for the sludge conditioned on the pilot-scale system has a slightly different shape. It rises very rapidly to the peak and then decays resulting in a curve that is more definitely skewed to the right than those generated on the batch system. This tended to be a characteristic of all the curves from the pilot-scale system. The amount of polymer required to produce a significant peak increased from 2 kg/t in Figure 6 to 4 kg/t in Figure 12. Since these tests were conducted almost a year apart, it is likely that changes in the sludge characteristics are responsible for the different polymer requirements. It is also evident that in the pilot-scale system the presence of a peak represents a necessary but not exclusive condition for optimum conditioning. Curve C, at 3 kg/t, shows a peak but the CST of 118 seconds does not indicate a well-conditioned sludge. It is necessary to add between 3.5 and 4.0 kg/t of polymer before a CST of less than 20 seconds is achieved. It can be concluded however, that the trends observed and the assumptions made regarding the batch data are generally valid for a continuous pilot-scale system.

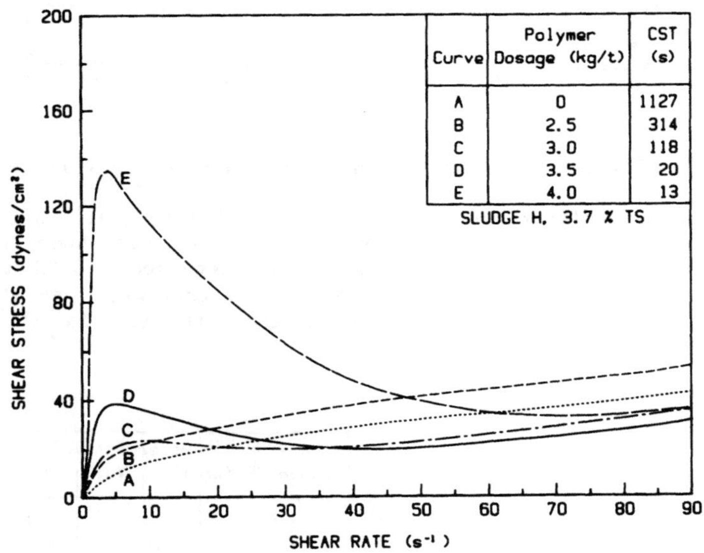

Figure 12. Effect of polymer dosage on rheology (pilot-scale).

In Figure 13 an attempt was made to show what effect varying the solids concentration has when the polymer flow rate is not adjusted accordingly. In this test the solids concentration was altered by adding dilution water but the total sludge flow rate (including dilution water) and the polymer flow rate were kept constant. Under these conditions, as the solids concentration decreases, the polymer dosage on a unit basis (kg/t) increases. At the highest solids concentration (4.4%) the effective polymer dosage is 3.2 kg/t and results in a CST of 37 seconds. At this CST the sludge would not be considered to be super flocculated, and while a peak is present in the rheogram, it is not substantial. As the solids concentration decreases to 3.1% the effective polymer dosage increases to 4.4 kg/t, the CST decreases to 17 seconds, and the rheogram peak becomes quite pronounced. The situation has changed from one where the polymer dosage is less than optimum to one where polymer overdosing may be occurring.

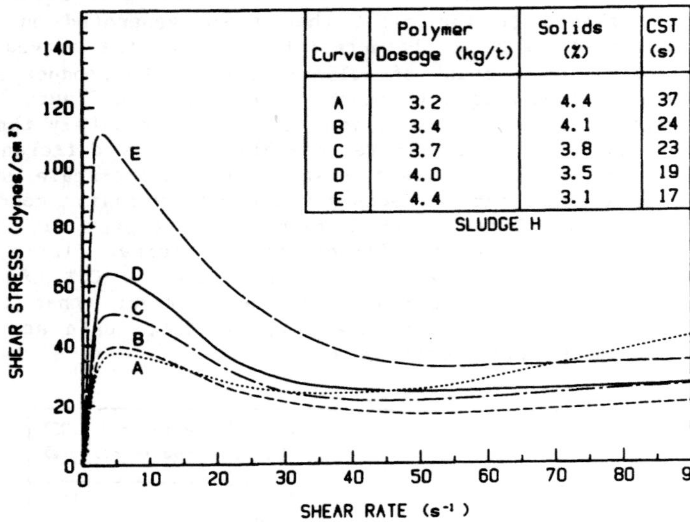

Figure 13. Effect of solids concentration on rheology (pilot-scale).

The final experiment looked at the effect of detergent and the data is shown in Figure 14. The results are identical to those shown in Figure 9 for the batch system. The addition of the detergent (0.2% by volume) increases the CST and destroys the peak in the rheogram. As the polymer dose increases, the CST begins to drop and the peak in the rheogram reappears.

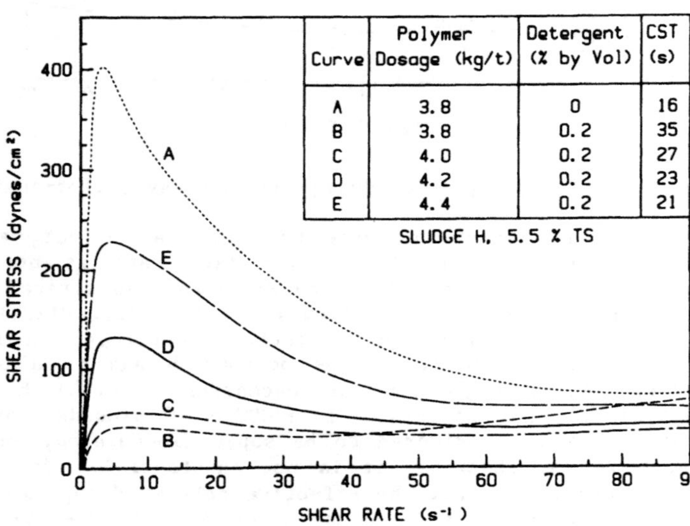

Figure 14. Effect of detergent addition on rheology (pilot-scale).

CONCLUSIONS

The work presented in this paper completes the series of experiments where test conditions were varied manually to evaluate fundamental relationships between polymer addition and rheological characteristics. It provides the necessary background for initiation of further work on the automatic control of sludge conditioning, based on rheological considerations. The major results of the study can be summarized as follows:

1. The viscometer equipment was upgraded and interfaced with a mini computer to provide the capability of generating the derivative of a sludge rheogram and of controlling the viscometer automatically.
2. Using a standardized test methodology, all anaerobically digested sludges exhibited a peak in the rheogram at some specific polymer dosage.
3. A peak in the sludge rheogram is necessary, but is not the only prerequisite for optimal conditioning in a continuous-flow system. This fact must be addressed when developing control algorithms.
4. Rheological properties are sensitive to, and provide a measurement of variations in the solids concentration or dewaterability of a sludge.
5. The relationships presented between polymer addition and rheological properties are not artifacts of batch mode, bench-scale systems. A continuous-flow pilot-scale sludge conditioning system and an in-line viscosity sensor were used to confirm the results established at bench-scale.

FUTURE WORK

The research activities in this program will continue to utilize the pilot-scale sludge conditioning system and will include computer control of the sludge, dilution water and detergent pumps such that variations in solids concentration and sludge dewaterability can be programmed as a function of time. The rheological properties of the conditioned sludge will be determined at regular time intervals and the data logged in the computer. Using control algorithms based on the relationships between rheological properties and polymer dosage, the computer will automatically make the appropriate adjustments to the polymer feed pump. Assuming this control strategy is successful at pilot-scale, tentative plans have been made to demonstrate automatic control of polymer conditioning for a full-scale belt press in 1984.

The other area that will be actively investigated is the requirements of the viscosity measurement system. A viscometer such as the RV-100, which is capable of generating a complete flow curve, is an expensive instrument and has features which are desirable for research, but are probably not necessary for a field application. If the relationship between shear stress and time, at a single fixed shear rate, can be correlated to the polymer dosage, the potential exists for using a viscometer which is less expensive and less complicated.

REFERENCES

1. Campbell, H.W. and B.P. LeClair (1975). "Dewatering Base Metal Mine Drainage Sludge", presented at the 10th Canadian Symposium on Water Pollution Research, Toronto.

2. Campbell, H.W., R.J. Rush and R. Tew (1978). "Sludge Dewatering Design Manual", Canada-Ontario Agreement Research Report No. 72.

3. Campbell, H.W., and P.J. Crescuolo (1982). "The Use of Rheology for Sludge Characterization", Water Science Technology, Vol. 14 Capetown, pp. 475-489.

4. Cost-Project 68 (1975). Sewage Sludge Processing, Commission of the European Communities Report No. EUCO/SP/48/75.

5. Dick, R.I. (1979). "Physical Properties of Activated Sludge", presented at the NATO Advanced Study Institute, Sludge Characteristics and Behaviour, University of Delaware.

6. Rimkus, R.R. and R.W. Heil (1975). "The Rheology of Plastic Sewage Sludge", Proceedings of the Second National Conference on Complete Water Reuse, Chicago, American Institute of Chemical Engineers, L.K. Cecil, ed.

7. Wastewater Technology Centre (1975). Unpublished data.

DISCUSSION

J. Eveson

Would it be correct to say that the method did not work for sludge G? Were all the sludges from plants having a phosphorus precipitation stage?

H.W. Campbell

The method worked less well for sludge G, which may have been due to the use of a non-optimal polymer - the polymer dose required for sludge G was about twice that required for the other two sludges. However, the rheogram for sludge G at a polymer dose of 4 kg/t does exhibit a peak and at this dose, its CST is reduced to 20 seconds.

The sludges used in the tests were all anaerobically digested sludges from plants where phosphorus precipitation was being carried out.

B. Paulsrud

Do you think that the rheological method could be used to control conditioning for dewatering by centrifuge?

Yes, I believe it could. The reasons our work is linked to belt presses are that the data we are generating is suitable for use on belt presses, belt presses are currently very popular in North America and they are easy to sample. It may not be so easy to get sludge samples from a filter press or a centrifuge. However, it should be possible to continuously monitor a small side-stream and through this optimise polymer addition.

P. Torpey

Could you comment further on the relation between solids concentration and rheology. Does the dependence of rheological behaviour on solids concentration mean that the latter parameter must be measured to provide an input to the algorithm for the control of polymer dose level?

Attempts have been made to control polymer dosing through solids concentration measurement but the problem is that no satisfactory sensor for solids measurement in the concentration range 2 to 7%, exists. It is true that rheological behaviour varies with solids concentration but this does not necessarily mean that solids concentration has to be measured for polymer dosing control. If, for instance, the polymer dose is increased to produce a peak in the rheogram, it does not matter whether the required increase is due to a change in the sludge solids concentration or its dewatering characteristics or both.

P. Balmer

Did you require any special pretreatment of the sludge for the continuous flow viscometry work, so as to prevent jamming of the viscometer by large particles?

No pretreatment of the sludge, other than use of an in-line mixer, was used. The performance of the system in this respect was better than expected. We were, of course, using digested sludge. I do not think the procedure would be satisfactory for raw sludge and I could see potential problems also at full scale.

L. Spinosa

How do you propose to control polymer overdose as the peak on the rheogram appears to increase with polymer dose?

Control is difficult to achieve at bench scale but should not be so difficult at pilot plant or full scale. For example, we could gradually increase the polymer dose until a peak appeared on the rheogram or until other shape properties of the rheogram were matched.

R.C. Frost

You suggested that the rheogram shapes could be partly explained by thixotropic effects. Do you think that particle migration and solids/liquid separation effects also exert an influence on the rheological behaviour? Could the peak shear stress be due to jamming of large particles in the viscometer gap?

The peak effect is clearly not due to particle-jamming as it can be also observed when the viscometer rotor is used on its own. In fact, we are considering using a star-shaped rotor on its own and simply measuring the torque instead of the shear stress. We have not investigated the phenomenon of particle separation. We get very good reproducibility at low levels of conditioning, while reproducibility is more problematic for well-conditioned sludges, due mainly to the difficulty of sampling the latter. We are satisfied from our experience that consistent results can be obtained for theological measurements of digested sludge. We have not, as yet, carried out any detailed mathematical analysis of the rheograms.

G. Hoyland

Is dosing of polymer at a fixed proportion to your solids concentration the object of your control system or how is the optimum dose determined?

The aim is to dose at an optimum rate as defined by the rheogram, which in turn is related to a target optimal CST value. For belt-press dewatering, we would aim at a CST of about 20 seconds. We could use the peak as an identifying characteristic of the required rheogram but we are working towards the inclusion of additional shape properties.

F. Colin

From a theoretical point of view, your rheograms are difficult to interpret. At shear rates beyond the peak value, the rheograms have negative slopes, indicating negative viscosity values. What does a negative viscosity mean?

I do not know what the meaning of the negative graph slope is in viscosity terms. There is an observed increase in shear resistance up to a peak value, followed by a decreasing shear resistance in the downpart of the rheogram. It would appear

I think there is another parameter which is of great importance in the time following polymer addition in batch experiments. We have developed a method of assessing conditionability of a sludge, based on the observation of torque versus time at constant rotational speed - we use a special device instead of a cylinder to prevent solids/water separation.

that this decrease is due to a breakdown of the sludge structure at the high shear rates.

DESIGN PROCEDURE FOR BATCH CONSOLIDATION OF SEWAGE SLUDGES

G HOYLAND, M DAY
Water Research Centre

SUMMARY

The paper describes a design procedure for batch consolidation of sewage sludges, probably the most cost-effective process to be found at sewage treatment works. For any particular sludge, tests are performed using a small centrifuge and perhaps a pilot-scale thickener to evaluate three parameters that characterise the sludge's consolidation properties. Given the parameter values, the performance of full scale consolidation tanks of any shape and size, with or without a mechanical thickening aid, can be predicted from a mathematical model.

INTRODUCTION

Batch consolidation, when used efficiently, is probably the most cost-effective process to be found at sewage treatment works. Used to reduce sludge volume, the process can reduce the cost of subsequent treatment and disposal by factors of two and more. Also, the process is the most common, occurring at least once at most works, and probably the simplest, requiring no more than sludge to be left perhaps with gentle stirring in a tank for a few days. However in practice, many consolidation tanks are poorly designed and experience operational problems. For example, it is not uncommon to find sludge floating at the top of a consolidation tank rather than consolidating at the bottom.

This paper describes a procedure, developed at the Water Research Centre, for designing batch consolidation tanks. Essentially, the procedure consists of performing tests on the particular sludge, using a small centrifuge and perhaps a pilot-scale thickener, to evaluate three sludge characteristics comprising the compression coefficient, the compression index and the resistance to consolidation. Given the parameter values, the shape and size of the prospective consolidation tank can be determined for a given duty.

THEORY

The basis of the design procedure is a mathematical model of consolidation developed by Hoyland(1). Although a detailed description of the model is outside the scope of this paper, the underlying assumptions in the model are described.

The model assumes that sewage sludges are comprised of individual particles or flocs suspended in water, and each particle is taken to be composed of a matrix of the suspended solids with physically-bound but displaceable water in the interstices. Adjacent particles are in contact, which distinguishes consolidation from settlement, the other form of gravity thickening. As depicted in Figure 1, the particles descend under their own apparent weight displacing water upwards. Since adjacent particles are touching the apparent weight of the particles is transmitted down the thickener so that the particles at the bottom are under the highest compressive pressure. Assuming acceleration forces are insignificant, the difference between the apparent weight of the particles above any particular elevation and the hydraulic friction acting on those particles is counterpoised by the compressive pressure acting on the particles at the particular elevation. Under the compressive pressure the particles deform such that the volume of the particles and the interstitial volume between the particles reduces, causing water to be displaced from both within and between the particles.

The concept of bound water(2) accounts for the observation that the concentration of solids in a sewage sludge at the transition between settlement and consolidation, called the compression point, is typically about 10 to 20 kg/m^3, equivalent to a volume fraction of about 0.006 to 0.012. However, assuming the particles at the compression point are spherical and randomly packed, they will have a volumetric concentration of at least 0.6, about 50 to 100 times greater than the volume of the solids. Bound water makes up the difference in volume and at the compression point must account for at least 98% of particle volume. Even at the high concentrations occurring at the bottom of thickeners, bound water probably account for about 95% of particle volume.

Perhaps the most fundamental assumption made in the model is that the specific resistance to filtration of the sludge blanket at any elevation in the consolidation tank is taken to be linearly related to the local compressive pressure. Such a relationship is normally found for high pressure filtration through sewage cakes and appears to be valid in consolidation even though the compressive pressures between the two processes differ by orders of magnitude.

SOLUTION

The thickening process described above has been modelled in terms of partial differential equations wrt time and distance, similar to those described by Shirato(3). Solutions to the model invariably give an excellent fit to the following equation:

$$D/(1-D) = T/T_{50} \qquad (1)$$

where D is called the degree of completion
T_{50} is the time (s) required to reach a degree of completion of 0.5
and T is time (s)

The degree of completion is defined by

$$D = \frac{C - C_o}{C_\infty - C_o} \qquad (2)$$

where C_o is the initial concentration (kg/m^3) of suspended solids in the sludge
C is the average concentration (kg/m^3) of solids in the sludge blanket at any particular time
and C_∞ is the ultimate concentration (kg/m^3) which would occur if thickening continued for an infinite time.

At the end of a consolidation flow ceases and the compressive pressure at the bottom of the tank reaches its ultimate value equal to the apparent weight of the suspended solids. This ultimate pressure determines, to a good approximation the ultimate concentration (C_∞). An empirical relationship between the ultimate pressure and concentration expressed as mass ratio of solids to water is

$$H_o C_o G(1- \frac{\rho_w}{\rho_s}) \times 10^{-2} = \frac{W_\infty^n - W_p^n}{W_{100}^n - W_p^n} \qquad (3)$$

where H_o is the initial height (m) of sludge in the consolidation tank
G is G-force (m^2/s)
ρ_w is density of water (kg/m^3)
ρ_s is density of solids (kg/m^3)
W_∞ is the ultimate mass ratio of suspended solids to water
W_p is the mass ratio found at the compression point
W_{100} is the mass ratio corresponding to an ultimate compressive pressure of 100 N/m^2
and n is a constant known as the compression index of the particular sludge

In accordance with intuition, the relationship implies that a sludge under no compressive force would be at its compression point and a sludge under an infinite compressive force would be completely dry.

The relationship between W and C is given by

$$W = \frac{C}{\rho_s - C} \cdot \frac{\rho_s}{\rho_w} \qquad (4)$$

which, for the range of concentrations found in sewage sludge may be simplified to

$$W \quad \frac{C}{\rho_w} . \qquad (5)$$

Substituting equation 5 into 3 gives

$$H_o C_o G(1 - \frac{\rho_w}{\rho_s}) \times 10^{-2} = \frac{C_\infty^n - C_p^n}{C_{100}^n - C_p^n} \qquad (6)$$

where C_p is the concentration of suspended solids (kg/m^3) found at the compression point
and C_{100} is the concentration (kg/m^3) corresponding to an ultimate compressive pressure of 100 N/m^2.

Because the value of the index (n) is usually in the range about 3 to 5 Equation 6 can be simplified giving

$$H_oC_oG\left(1-\frac{\rho_w}{\rho_s}\right)\times 10^{-2}=\left(\frac{C_\infty}{C_{100}}\right)^n \qquad (7)$$

When dealing with any particular sludge, it is expedient to simplify further by incorporating the density term and C_{100} into a single constant, i.e.

$$H_oC_oG\times 10^{-2}=\left(\frac{C_\infty}{K}\right)^n \qquad (8)$$

where K, called the compressive coefficient (kg/m^3) is given by

$$K = C_{100}\left(1-\frac{\rho_w}{\rho_s}\right)^{1/n} \qquad (9)$$

Factors Affecting T_{50}

The value of T_{50} in Equation 1 is determined by the shape of the consolidation tank and the thickening properties of the sludge. Solutions to the model supported by experimental evidence suggest(1) that, to a good approximation

$$T_{50} = \frac{\mu R}{K^n}\cdot\frac{C_o^{\,n-1}(C_\infty - C_o)H_o}{G} \qquad (10)$$

where μ is the viscosity (N.s/m^2) of water. The characteristic R has units of m/kg, the same as specific resistance to filtration. Indeed, R is a measure of (but not equal to) the specific resistance to filtration of the sludge blanket in the consolidation tank and is therefore called the specific resistance to consolidation. As discussed later, the value of R depends on the scale of the consolidation tank and the efficiency of any mechanical thickening aid, present in the tank, as well as the nature of the sludge.

An implication of Equations 1 and 10 is that, if the compressive pressure between consolidations is constant, the time required to thicken any particular sludge of a given concentration to any particular degree of completion is inversely proportional to the square of the G force, i.e.

$$T \propto G^{-2}\big|_{D,H_oG\,C_o} \qquad (11)$$

Unlike Equation 10, this proportionality which is derived directly from the model is exact and can be exploited when performing centrifuge tests.

DESIGN PROCEDURE

The design procedure consists essentially of using Equations 1 and 10 to evaluate the characteristics, K, n and R from results of tests performed using a small centrifuge and a pilot-scale consolidation tank. Having evaluated the characteristics the performance of the prospective full-scale tank can be predicted depending on its shape and size.

HARDWARE

The centrifuge and pilot-scale consolidation tank are described as follows:

Centrifuge

The centrifuge is a standard instrument available from Triton Electronics Ltd, Bigods Hall, Dunmow, Essex, England. Figure 2 is an outline of the centrifuge[4] showing the rotating arm which carries two pivotted flat-bottomed tubes, 90 mm long x 22 mm diameter. The two tubes allow duplicate measurements to be made as well as balance the arm. G-forces up to about 1000 m/s^2 (100 g) can be simulated by rotating the arm at controlled speeds of up to 800 rpm, and sludge may be placed in the tubes to depths of between 4 and 8 cm. When using sludge depths less than about 4 cm, changes in the depth of the sludge blanket in the tube are small and cannot be accurately measured. A stroboscope synchronised with the arm allows the progress of the consolidation in either tube to be viewed through a window in the centrifuge body, and measurements should be continued until the sludge/water interfaces are stationary.

Pilot-Scale Thickener

A pilot-scale consolidation tank is not yet commercially available. Figure 3 shows that the tank used at the Water Research Centre is essentially a transparent plastic column, 150 mm diameter x 2 m high. A picket fence comprising two vertical rods, 13 mm diameter, supported by a central shaft can be rotated in the tube to assist consolidation. To use the column, a sample of the sludge is poured in the top and allowed to consolidate with or without the fence rotating as required. Progress of the consolidation is followed by viewing the top of the sludge blanket through the plastic.

EVALUATION OF CHARACTERISTICS

The evaluation of the consolidation characteristics is exemplified with tests performed with the centrifuge and pilot-scale tank on samples of a primary sludge taken from a large operational sewage treatment works. The values of the consolidation resistance predicted from the tests are compared with the value measured in a full-scale tank.

Compression Characteristics

To evaluate the compression characteristics several centrifuge tests are performed over a wide range of G-forces. Height of sludge in the centrifuge tube may also be varied. Each test should be continued until the sludge/liquid interface remains stationary for about 1 hour indicating that the consolidation is complete and the solids are at their ultimate concentration. If the thickening is terminated early, the value of C_∞ can be estimated by plotting $1/(C-C_0)$ against $1/T$ and extrapolating to $1/T=0$, as shown in Figure 4.

Plotting the measured values of C_∞ against H_0C_0G on logarithmic axes, as shown in Figure 5, should give a straight line in accordance with Equation 8. The gradient of the line defines the value of the compression index, n, and its intercept at the compressive pressure of 100 N/m^2 the value of the coefficient, K.

Resistance to Consolidation

Evaluation of R is comparatively involved because the value of the characteristic can depend[5] on tank diameter and the presence in the tank of mechanical thickening aids such as picket fences. Thickening aids decrease resistance by promoting flocculation within the sludge, and small tanks increase resistance by suppressing flocculation. Suppression, however, is confined to tanks less than about 50 mm diameter where tank and floc sizes are comparable. Because centrifuge tests are performed in 22 mm diameter tubes without a mechanical aid, resistance values derived from the results can be high and must be divided by a correction factor, called the enhancement ratio, as follows

$$R_f = R_c/E \tag{12}$$

where R_f is the resistance relating to full scale
R_c is the resistance derived from results of a centrifuge test
and E is the enhancement ratio.

In practice, the value of E depends on the degree to which the sludge flocculates naturally and on the efficiency of the mechanical thickening aid. Results from an evaluation[5] of the consolidation of primary and activated-sludge thickening with and without picket fences suggests that the value of E averages about 3 for unassisted thickening and about 9 for thickening assisted by a well designed and operated picket fence. However, the value of E for assisted thickening can vary between sludges from unity, for those which show no tendency to flocculate, up to 20. Confidence in the accuracy of resistances derived solely from the results of centrifuge tests is therefore low. A reliable value of resistance can, however, be derived by performing a test with the pilot scale consolidation tank with or without the picket fence rotating as appropriate. Because the diameter of the pilot-scale tank is greater than 50 mm, the resistance will be similar to that in the full-scale tank and no correction need be applied.

To evaluate resistance using a centrifuge, a test is performed at a compressive pressure similar to that in the prospective full-scale tank, and the results plotted as shown in Figure 6. In accordance with Equation 1, the plot of D/(1-D) against time gives a straight line passing through the origin. The resistance can be conveniently calculated by noting the time to 75% completion, T_{75}, and substituting the value into the following expression derived from Equations 1 and 10:

$$R = \frac{T_{75} G\, K^n}{3 \mu H_o C_o^{n-1} \, (C_\infty - C_o)} \tag{13}$$

Figure 6 also shows similar plots for the results of a test performed in the pilot-scale tank, with the picket fence operating in a sludge depth of 1.6 m, and of a full-scale consolidation in a 4.2 m diameter x 3.2 m high tank also with a picket fence. As shown in Table 1, the T_{75}'s from the plots give resistances of 0.55 Gm/kg for the full-scale and 0.46 Gm/kg for the pilot-scale, and these can be compared with the value of 0.90 Gm/kg derived by dividing the resistance of 8.2 Gm/kg derived from the centrifuge test by the average enhancement ratio of 9. In this particular case therefore, the pilot-scale thickening under-estimated the resistance measured in the full-scale tank by only 6% and the centrifuge over-estimated by about 70%.

Table 1. Calculation of consolidation resistances from results of centrifuge, pilot-scale and full-scale tests

Equipment	T_{75} (s)	H_o (m)	G (m/s^2)	μ (Ns/m^2)	C_∞ (kg/m^3)	R (Gm/kg)
Centrifuge	2.2×10^2	0.052	600	0.001	73	8.2
Pilot-scale	2.4×10^4	1.6	9.8	0.0013	60	0.50
Full-scale	6.8×10^4	3.2	9.8	0.0013	72	0.53

Notes:

(i) Compression index, n = 4.1

(ii) Compression coefficient, K = 47 kg/m^3

(iii) $$R = \frac{T_{75}\, G\, K^n}{3\mu H_o C_o^{n-1}(C_\infty - C_o)}$$

Generally, consolidation resistances derived from the results of pilot-scale thickenings are within 50% of the 'full-scale' values. In contrast, confidence in values predicted from centrifuge tests is comparatively low.

The curvature in the plot for the pilot scale thickening in Figure 6 is symptomatic of anaerobic or anoxic activity. Bubbles of methane or nitrogen produced by the activity bouy the suspended solids, causing the ultimate concentration, C_∞, to progressively decline. This behaviour is normally observed when the consolidation is quiescent especially when temperature is above about 12°C. Indeed, most raw primary sludges will float if held under these conditions for more than a day or two. A picket fence turning in a thickener prevents flotation, in addition to promoting flocculation, by releasing most of the bubbles, although some bubbles can remain and reduce the compressive pressure.

PERFORMANCE ON FULL-SCALE

Having evaluated the characteristics n, K and R, the effect of tank volume and shape on performance can be predicted. Equation 1, which models the performance of the prospective consolidation tank, can be expressed in terms of volume, V (m^3), of sludge in the tank as follows:

$$\frac{D}{1-D} = \frac{V\, C_o}{M\, T_{50}} \tag{14}$$

where M is the time-averaged mass flowrate (kg/s) of suspended solids through the tank. When designing a tank for a particular site, M is usually constant and may therefore be combined with V to give a unit volume, V_u (m^3 s/kg), i.e.

$$\frac{D}{1-D} = \frac{V_u C_o}{T_{50}} \tag{15}$$

Substituting for D and T_{50} from Equations 2 and 10 defines the final concentration, C, in terms of the operational parameters, as follows:

$$\frac{C - C_o}{C_\infty - C} = \frac{V_u G K^n}{\mu R C_o^{n-2}(C_\infty - C_o)H_o} \tag{16}$$

Further, a specific volume V_s (mm^2/s) can be defined by combining V_u and R, i.e.

$$V_s = V_u/R \tag{17}$$

such that

$$\frac{C - C_o}{C_\infty - C} = \frac{V_s G K^n}{\mu C_o^{n-2} (C_\infty - C_o)H_o} . \tag{18}$$

The relationship between specific volume and retention time is given by

$$V_s = \frac{T}{RC_o} . \tag{19}$$

Implications

The implications of Equation 18 in design are exemplified by considering the full-scale consolidation of a primary sludge having characteristics similar to those given in Table 1, i.e.

Characteristic	Value
Compression index, n	4.0
Compression coefficient, K	50 kg/m^3

The prospective consolidation tank succeeds primary sedimentation and the concentration of suspended solids in the sludge withdrawn from the primary sedimentation tank can be varied in accordance with the method of withdrawal. The effect of initial concentration on the performance of the consolidation tank therefore requires investigation as well as the effect of tank shape. To release bubbles produced by biodegradation and to promote flocculation the tank is fitted with a picket fence.

Solutions to Equation 18 are illustrated for values of specific volume equal to 2, 10 and 50 mm^2/s which covers the range normally found in practice. In accordance with Equation 19, the low value of specific volume represents a low retention time and/or a high consolidation resistance, and conversely, the high value of V_s a high retention time and/or low resistance.

Figures 7 to 9, illustrating the solutions, show final concentrations plotted as contours between axes representing tank heights from 1 to 10 m and initial concentrations from 20 to 90 kg/m^3. Of this range, initial concentrations up to about 60 kg/m^3 only are of any practical interest given the limitations to consolidation in primary sedimentation tanks.

Figure 7 pertaining to a V_s value of 2 mm^2/s shows, that, for any particular initial concentration, there is an optimum tank height which maximises the final concentration. If the sludge has an initial concentration of 50 kg/m^3, the optimum tank height will be less than 1 m giving a final concentration of about 65 kg/m^3. Figure 8 shows that a V_s value of 10 mm^2/s increases optimum tank height as well as final concentration. For sludge at an initial concentration of 50 kg/m^3, optimum height would be about 2 m and final concentration about 77 kg/m^3. Note that doubling the initial concentration from 30 to 60 kg/m^3 increases final concentration (at the optimum height) from about 73 to only 82 kg/m^3. Optimum tank heights in Figure 9, pertaining to a V_s value of 50 mm^2/s, range from 10 m to 5 m. For an initial concentration of 50 kg/m^3 the optimum height would be about 6 m giving a final concentration of about 96 kg/m^3. In practice, however, this performance may not be realised given that long retention times can allow bubbles, produced by biodegradation, to collect in the sludge and destroy the compressive pressure.

CONCLUSIONS

A procedure for designing batch consolidation tanks for thickening sewage sludges has been described. Essentially, the procedure consists of performing tests on the particular sewage using a small centrifuge and perhaps a pilot-scale tank, and processing the results with the aid of a mathematical model to evaluate three parameters. Two of the parameters, characterising the compression properties of the sludge, are evaluated from centrifuge tests, and the third, characterising the permeability, is most accurately evaluated from a test with the pilot-scale tank. The procedure has been exemplified with tests performed on a primary sludge.

Predictions made by the model indicate that depending on the size of the consolidation tank and the specific resistance to consolidation of the sludge there is an optimum tank height which maximises the concentration of solids in the consolidated sludge. Optimum height ranges from 1 m or less, for tanks with small unit volumes or those containing sludges with high specific resistances, up to about 10 m for tanks with large unit volumes or those containing sludges with low consolidation resistances.

NOMENCLATURE

C	Average concentration of suspended solids (kg/m^3), in consolidation tank at any particular time or final concentration
C_o	Initial concentration of suspended solids (kg/m^3)
C_∞	Ultimate concentration of suspended solids (kg/m^3)
C_p	Concentration of suspended solids (kg/m^3) at compression point
C_{100}	Concentration of suspended solids (kg/m^3) in a sludge under a compressive pressure of 100 N/m^2
D	Degree of completion
E	Enhancement ratio

G	Force due to acceleration (m/s^2)
H_o	Height of sludge in consolidation tank
K	Compression coefficient (kg/m^3)
M	Time-averaged mass flowrate of suspended solids (kg/s)
n	Compression index
R	Resistance to consolidation (m/kg)
R_f	Resistance to consolidation (m/kg) in full-scale tank
R_c	Resistance to consolidation (m/kg) in centrifuge tube
T	Retention time of sludge (s)
T_{50}	Retention time (s) required for 50% completion
V	Volume (m^3) of consolidation tank
V_u	Unit volume ($m^3.s/kg$) of consolidation tank
V_s	Specific volume (m^2/s)
W	Mass ratio of suspended solids to liquid in sludge
W_∞	Ultimate mass ratio of suspended solids to liquid in sludge
W_p	Mass ratio of suspended solids to liquid in sludge at its compression point
W_{100}	Mass ratio of suspended solids to liquids in sludge under a compressive pressure of 100 N/m^2
μ	Viscosity of water ($N.s/m^2$)
ρ_w	Density of water (kg/m^3)
ρ_s	Density of suspended solids (kg/m^3)

REFERENCES

1. Hoyland, G., Day, M.C. 'Consolidation of Sewage Sludges', WRC publication (in preparation).
2. Anderson, A.S., Hanson, R.C. 'A Simulation Model of the Gravity Separation of Concentrated Solids', Adv. Wat. Pollut. Res. Proc., 6th Int. Conf., Jerusalem 1972, Pergammon Press 1973, 457-474.
3. Shirato, M. et al. 'Analysis of Settling of Thick Slurries due to Consolidation', Journal of Chem. Eng. of Japan, 1970/71, 98-104.
4. Lockyear, C.F., White, M.J.D. 'The WRC Thickenability Test Using a Low-Speed Centrifuge, WRC publication TR118, May 1979.
5. Hoyland, G., Day, M.C. 'Evaluation of Picket Fences in Consolidation of Sewage Sludge', WRC publication (in preparation).

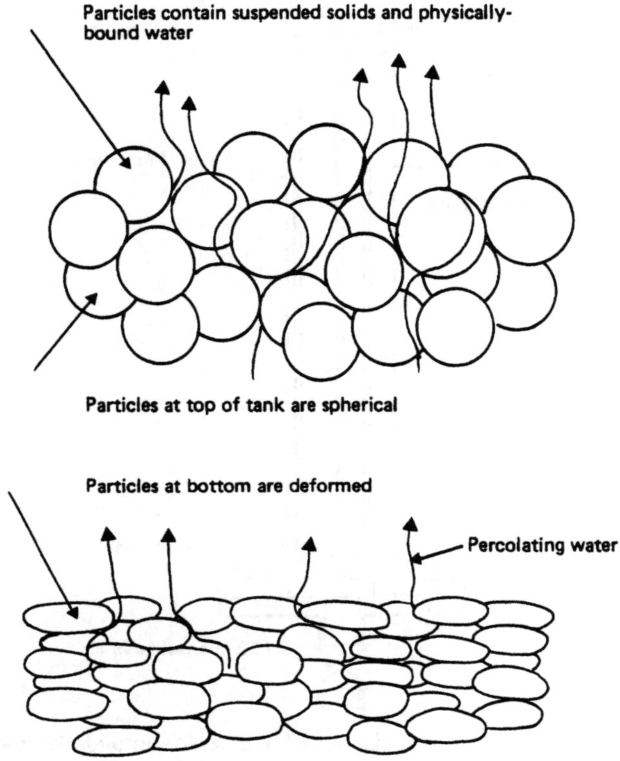

Fig. 1. Mechanism of consolidation

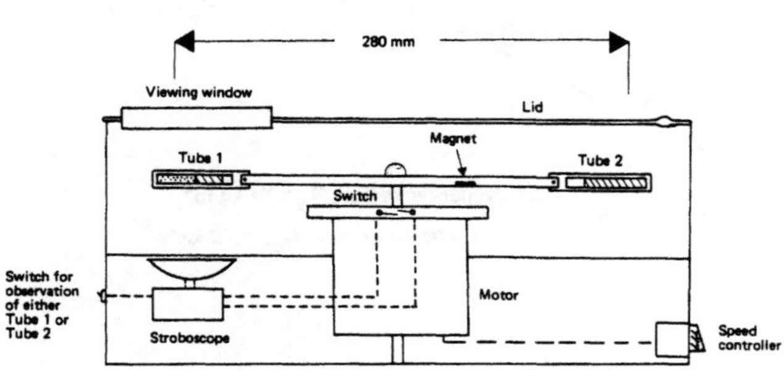

Fig. 2. Outline diagram of centrifuge

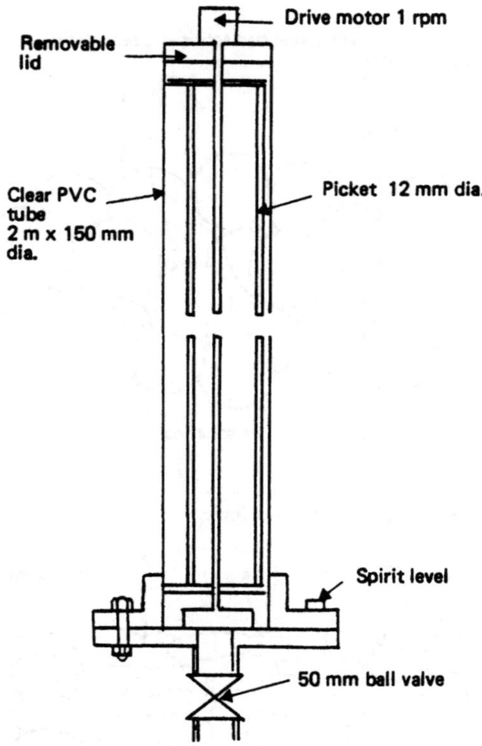

Fig. 3. Outline diagram of pilot-scale consolidation tank

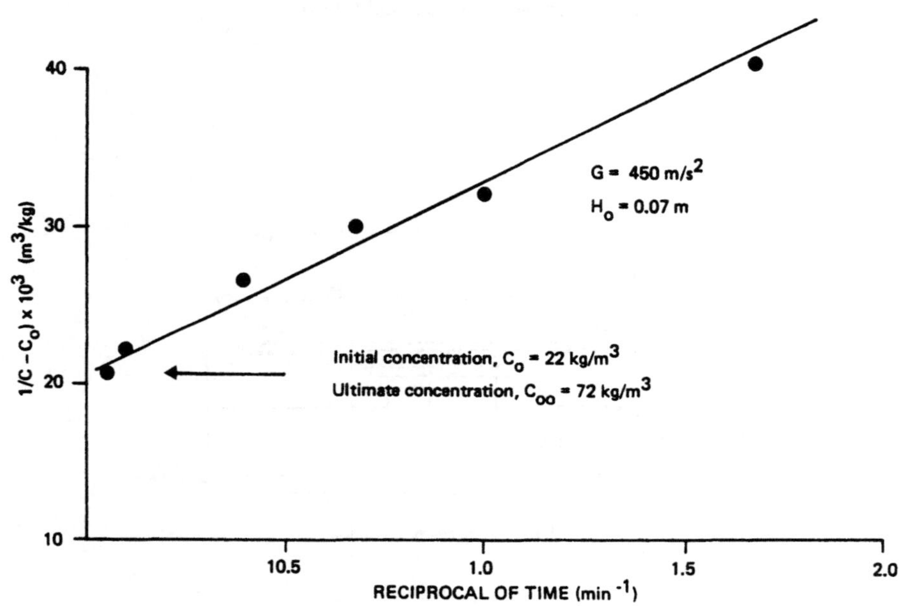

Fig. 4. Determination of ultimate concentration

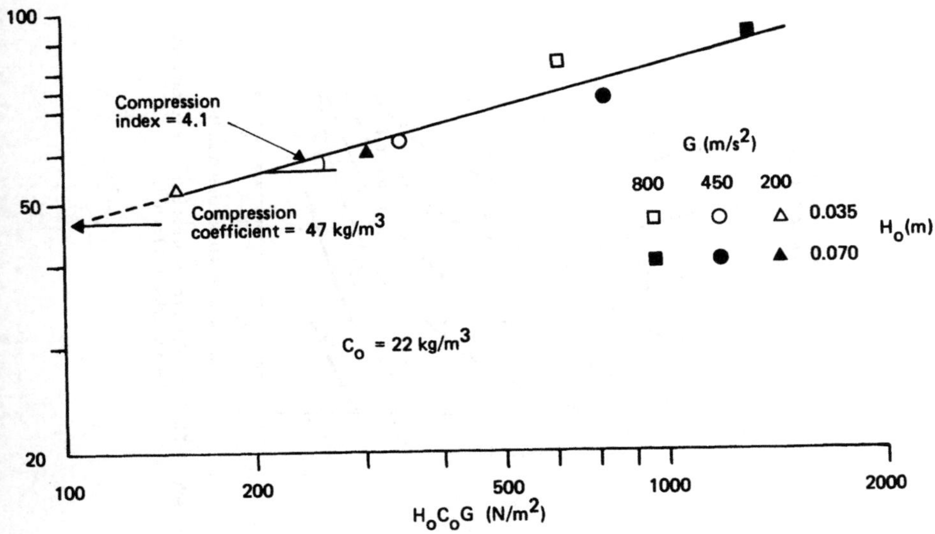

Fig. 5. Determination of compression characteristics

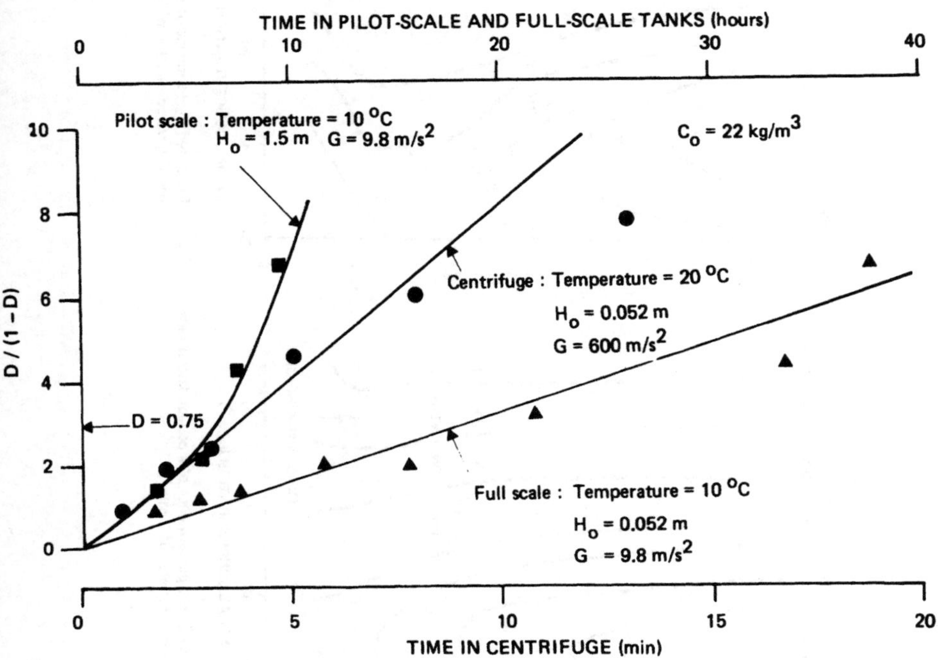

Fig. 6. Determination of specific resistance to consolidation

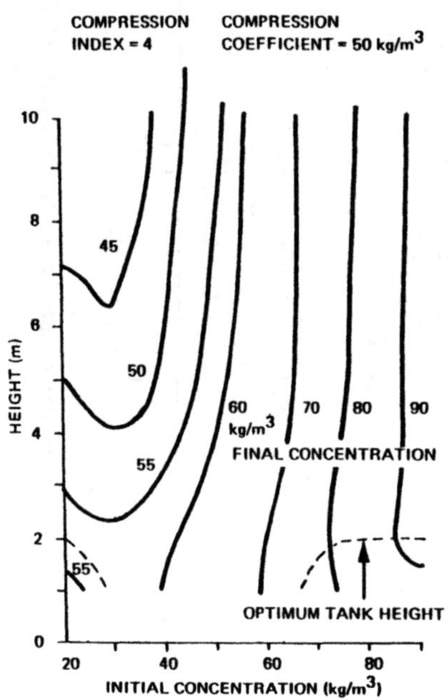

Fig. 7. Predicted performance of a consolidation tank for a V_s value of 2 mm^2/s

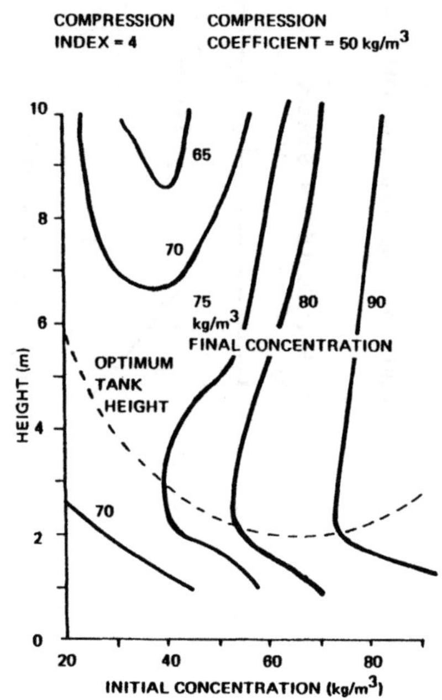

Fig. 8. Predicted performance of a consolidation tank for a V_s value of 10 mm^2/s

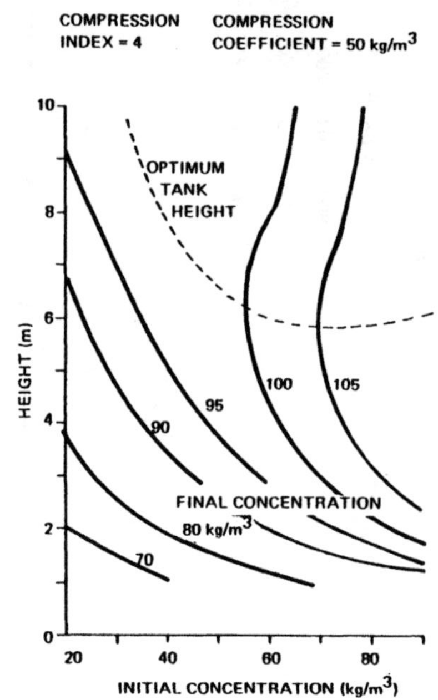

Fig. 9. Predicted performance of a consolidation tank for a V_s value of 50 mm^2/s

DISCUSSION

T. Casey

Dr. Hoyland's paper is a welcome contribution for a number of reasons. Firstly, despite the fact that thickening is a widely used and apparently simple process, the fundamentals of the process are not very well understood and, as a result, design knowledge is lacking. Secondly, I think it is true to say that this is the first time this aspect of sludge processing has been addressed in this working party.

Dr. Hoyland's approach is based on a 3-parameter 2-phase model and as such it is probably complicated enough to put many design engineers off examining it more closely. Unfortunately, however, the real situation may be even more complex still since in my experience it is the presence of a third phase - a gas phase - which very often plays the dominant role in sludge thickening. I am sure the phenomenon of sludge flotation in thickening tanks is a familiar problem to many in the audience. The presence of gas bubbles, even if not present in sufficient volume to cause flotation, greatly disrupts the thickening process since the latter relies entirely on the density difference between the solids and water fractions to generate the hydraulic gradient necessary to cause upward water movement in the consolidating sludge. I had occasion to observe a very striking example of this phenomenon recently when we were engaged in a pilot plant study of anaerobic digestion at the Ringsend STP here in Dublin. Within 24 hours

G. Hoyland

Thickening data from centrifuge tests, as plotted in Fig. 6, typically produces straight line plots, although small recording errors near test completion, may appear to give large deviations. Pilot plant and full scale thickening data, on the other hand, may plot non-linearly due to the influence of gas bubbles in the sludge. Gas bubbles reduce the available compression force and if present in excessive amount, may cause flotation. We have carried out extensive testing of picket fences and are satisfied that they effect release of gas bubbles during thickening. Sludge should be thickened as early as possible to avoid septicity and the consequent generation of gas.

Your comment on the thickening time relation between pilot plan and full scale is correct. The reason the plotted data on Fig. 6 does not appear to be in accord with this relationship is that the relevant time scale has been inadvertently omitted.

of the transfer of primary sludge from sedimentation tanks to thickening tanks, the entire mass of sludge floated in the latter. It is often suggested that the gas causing flotation is methane, but this could not be so in this instance as the pH of the sludge was in the range 5 to 6. My conclusion was that the gas was carbon dioxide and perhaps some hydrogen, resulting from acid fermentation in the raw sludge.

I would like now to refer to Fig. 6 in the paper, which shows graphical plots of thickening data from a centrifuge, a pilot unit and a full-scale thickener, presented in accordance with the model relationship given by equation (1). It seems to me that it would be more realistic to represent the centrifuge data by a curve rather than a straight line, as shown by the authors. Also, according to the model relationship, the ratio of the times required to reach the same D value in centrifuge and full scale should equal the inverse square of the respective G values. The plotted data does not seem to support this model relationship. Finally, I would like to put a practical question to the author. Are picket fences the best method of gas release from sludges and can they be satifactorily used in gravity-thickening of primary sludge?

<u>A.M. Bruce</u>

Can the batch thickening model be also applied to continuous thickening?

In principle, yes, but we have not extended it to continuous thickening.

CHARACTERIZATION OF ODOURS

N. Berg
Central Institute for Industrial Research
Oslo, Norway

Summary

Treatment of sewage and sludge is not odourfree operations and can therefore give rise to odour problems. When working with these odour problems and eventually odour abatement, two factors are of great interest to know:

1. The total odour strength.
2. Which organic components contribute to the total odour.

An olfactometer is most often used to measure the total odour strength, while the organic analysis is performed by use of a gas chromatographic technique. Results from investigations of odorous air collected at sewage plants can be summarized as follows:

1. Larger amounts of sludge normally give rise to the heaviest odour problems.
2. It is possible in advance to predict the strongest odour sources at the plants.
3. All kinds of odorous organic compounds may be found when air samples are analysed. The most important are the sulfur- and nitrogen-containing compounds.
4. It is possible with today's modern technology to minimize the odour problems by use of different reduction equipment.

1. INTRODUCTION

Collection and handling of wastewater and sludge have caused odour problems for many years. Odours have affected citizens and also been a problem for authorities, and the literature is full of refrains.

Already in 1931 a scientist said: "Odour problems are frequently one of operation, but recent developments have indicated that much of the difficulty can be anticipated and practically eliminated." (1)

Looking at today's situation, one clearly observe that the odour problems have not been eliminated. On the contrary, many people would postulate that the odour problems have instead increased.

The reason for this can perhaps be that new treatment plants are built close to populated areas, once isolated plants have been surrounded by new populations, plants are operated near or above maximum capacity, and last but not least, public have been more aware of odour problems and today no longer accept strong odour emissions from sewage treatment plants.

Everybody understand that sewage treatment plants are a necessity today. It must also be clear that wastewater and sludge are not odourfree substances, and because of this, treatment plants can not be compared with industries causing no odour emissions.

The first part of this paper will discuss how it is possible to measure and analyse odorous air samples, followed by a description of how it is possible to reduce odour problems at plants handling sewage and sludge.

2. HOW TO MEASURE ODOURS

When working with odour problems or odour reducing systems, two factors are of great interest to know:

1. The total odour strength.
2. Which organic components contribute to the total odour.

The total odour strength has to be measured by use of an olfactometric technique. The most common way to do this measurements today is to use an odour panel, smell the odorous air at different dilution levels, and then calculate the necessary dilution to reach odourfree conditions.

The analysis of odorous compounds has to be done by use of a chromatographic technique. Today the most commonly used method is gas chromatography.

2.1 Total odour strength.

There are several olfactometers available on the market. At our institute in Norway we have for 6 years used a dynamic triangle olfactometer developed by Dr. Dravnieks at IIT in Chicago (2). We know that many scientists, particularly in Europe, raise objections against Dr Dravnieks' olfactometer, and therefore have constructed their own instruments (3,4).

If the Dravnieks' olfactometer, however, is used with some caution, our opinion is that it works very well.

The olfactometer supplies 6 dilution levels. At each dilution level 3 samples ("triangle") are presented to the panelist (the total panel consists of 7-9 panelists) from a set of glass sniffing ports: two are test room air (blanks), and the third is the odorous gas diluted with the test room air. The panelist is instructed that one of the three ports in each set may exhibit an odour, and that his task is to smell the effluents from the ports and find which port, in his opinion, delivers an odorous sample. He must decide; if he feels that none of the three ports delivers an odour, he must simply make a guess.

The panelist proceeds from the most diluted sample towards higher concentrations of the sample. The choice is signalled by depressing a button corresponding to the port thought to be odorous, and this choice is observed by the panel leader on a panel of light in a separate signal box.

The panel leader records the judgements and calculates data following a statistical procedure which results in an averaged panel value termed ED_{50}. This term denotes Effective Dosage at the 50 percent level; it is that dilution at which 50 percent of the panel would and 50 percent would not detect odour of the diluted sample. The dilution is denoted by the dilution factor. For instance: ED_{50} = 1000 means that one litre of the odorous air must be diluted with 1000 litres of non-odorous air to reach the panel threshold termed ED_{50}.

With the olfactometer in use it is possible to measure dilutions between ca. 10x up to ca. 30.000x.

Experience has shown that the ED_{50} value measured in the laboratory can be transferred to the actual odour emission. This can be done among other things because the measurement takes place in a room completely free from foreign odours, and spesially tested, motivated panelists are used.

Air samples are collected in special plastic bags. The best bags have been shown to be made of Tedlar (polyvinylfluoride). After collection the bags can be stored for some time. Experiments with sewage odours have indicated that up to 48 hours storage do not influence the sample.

2.2 Analysis of odorous components.

When discussing instrumental analysis of odorous components, two important parameters must be remembered; the odour threshold value and the relationship between odour intensity and concentration.

Different chemical components have different odour threshold values. These threshold values can be measured in different ways, and large compilation tables can be found in the literature (5,6). Looking through the tables, it is easy to observe that the threshold value for one component may vary to a large extent. The reasons for this can be that different investigators have used different definitions for the threshold values and, therefore, interpreted their experimental results in different ways. These interpretations can vary from defining the threshold value as the lowest detection threshold to the highest recognition threshold.

The relationship between odour intensity and concentration of different organic compounds can be expressed as follows:

$$I = Kc^{n} \; ; \; \log I = \log K + n\log C$$

where

I = odour intensity
C = concentration of odour reaching the nose
k, n = coefficients, normally $0.2<n<0.8$.

By varying n, it can be seen that two samples at the same concentrations would show different intensities. Also the equation tells how difficult it can be to reach minimal odour intensity. Then the concentration has to be extremely reduced. If $n = 0.25$ and the concentration decreases down to 1/16 of what it was, the intensity would only decrease to half of its original value. This also means that if the concentration of the odour components are reduced about 90 %, there is a possibility that almost nothing has happened to the odour intensity.

The relationship between odour intensity and concentration is illustrated in figure 1.

When discussing odour problems at sewage treatment plants it can also be interesting to look at the threshold limit values (TLV) for different odorous compounds.

Table I gives the values for some less-odorous and odorous compounds. The interesting thing to remark is the difference between the threshold limit values for the most odorous compounds. Hydrogen sulfide, for instance, with a very low odour threshold, would exhibit strong odour although its concentration is far below the threshold limit value. Other sulfur-containing compounds behave in the same way. For the less-odorous compounds (acetic acid, benzene, ethanol) the opposite happens. The threshold limit values are reached long before one can smell the compounds.

As mentioned earlier, you only get a number, an ED_{50} value, when doing total odour strength measurements. The instrumental analyses, gas chromatography (GC), or gas chromatography/mass spectrometry (GC/MS) tell you, however, which different organic compounds that are present in your odorous sample and also the concentrations.

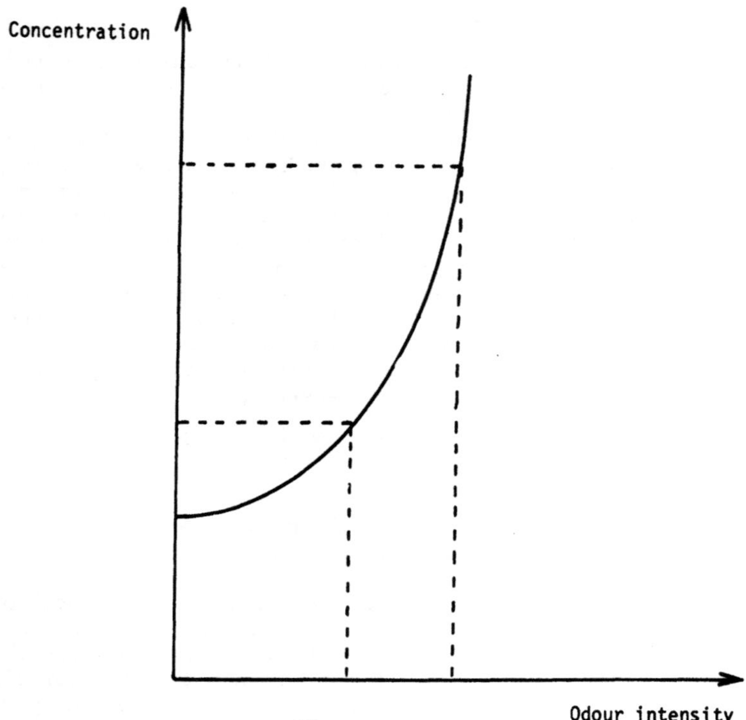

Figure 1.

Relationship between odour intensity and concentration.

Table I Odour threshold values and threshold limit values for some organic compounds. All values given as mg compound/m^3 air.

	Odour threshold	Threshold limit values*
Acetic acid	~ 3	25
Benzene	~ 15	3
Buturic acid	~ 0.05	?
Chlorine	~ 0.05	1.5
Ethanol	~ 200	1900
Hydrogen sulfide	~ 0.01	15
Mercaptans	~ 0.001	~ 1
Skatole	~ 0.0008	?

*Norwegian limit values 1981.

I do not want to go into details explaining the gas chromatograph. Only a very brief introduction will follow.

When doing gas chromatographic analyses, a sample, either the original one or a concentrated sample, is introduced (injected) into the chromatograph. The different components present in the sample then will separate and could later be detected and recorded. The results then appear in the gas chromatograms.

The separation of the different compounds present in the sample is dependent on several things such as the GC-oven temperature, the GC-column, and the carrier gas flow. There exist a lot of different detectors, some of them more universal like the most commonly used flame ionization detector, some of them more specific like the electron capture detector or the flame photometric detector used for sulfur compounds.

A special way to analyse odorous organic components is to use an "on-line" sniffing technique. After having passed through the column of the GC, about 80 % of the gas-stream is led to a sniffing device and 20 % to the detector. The person, performing the sniffing, gives comments which are correlated with the peaks on the chromatogram, and now it is possible to get a correlation between the amounts of the different compounds and the odour of the separated components. One interesting point here is that very often we observe that the human nose is a much better detector for odorous compounds than the electronic detector in the chromatograph.

The collection of an odorous sample and the preparations before the analyses very often cause much more trouble than the analyses itself. This happens because the odour threshold values of the components present in the sample often are much lower than the detection limit of the GC. This means that it is impossible to see anything on the chromatograms if an original, unconcentrated sample is analysed although the sample could show a very offensive odour. The only way to attain proper results when this happens is to concentrate the odorous compounds before the GC analysis.

This could be illustrated by an example. Sulfur compounds often show the problems mentioned above. Many sulfides and mercaptans have odour threshold values far below 1 ppm (parts per million), some of them below 1 ppb (parts per billion). The normal GC detection limit can be around 1 ppm or even ligher, and without a concentration step it would be impossible to analyse these kind of substances.

Many different concentration methods exist. When we talk about the analysis of odorous air samples, the use of solid adsorbents like activated carbon, porous polymers like Tenax, Chromosorb, and Porapack have been shown to be very effective. Some of the problems often found when using these materials are that one adsorbent can be very effective for concentration of one group of odorous compounds, but can show very bad capacity for adsorption of other groups of compounds. An universal adsorbent does not exist.

Investigations dealing with total analysis of all organic compounds present in air from treatment plants have been carried out several times (7,8,9,10,11). These investigations have shown that almost all groups of odorous compounds may be present. Table II shows some of the results.

Table II Odorous organic compounds found in air samples collected at sewage treatment plants.

Sulfur-containing compounds	Hydrogen sulfide, sulfides, mercaptans, thiophenes, thiazoles
Nitrogen-containing compounds	Amines, amides, pyrazines
Oxygen-containing compounds	Alcohols, aldehydes, ketones, acids, esters, phenols
Hydrocarbons (HC)	Unsaturated HC, aromatic HC

Together with these odorous compounds a lot of other organic components have been identified.

In most cases the heaviest odour problems are caused by sulfur- and nitrogen-containing compounds. These compounds have very low odour threshold values and in spite of very often being present in very low concentrations they produce heavy odour.

Hydrogen sulfide has been thought to be a very important odour constituent of sewage air. Recently performed investigations have, however, shown that this perhaps is not true. Instead of finding high H_2S concentrations some German scientists found several polysulfides, with even lower odour thresholds than H_2S, to cause odour problems in sewage systems (12). They also found that these polysulfides can be responsible for the corrosion of concrete sewage systems.

3. ODOUR SOURCES

It must be clear that odour problems could differ from one treatment plant to another. Although it sometimes can be very difficult in advance to predict how the odour problems would be, there are, however, possible to give some indications.

First of all, plant treating only wastewater most often do not give odour problems to the same extent as plants treating larger amounts of septic and sludge. Also it is possible to predict where at the plants the odour emissions would be strongest. Among such places can be mentioned:

- raw sludge hopper
- raw sludge holding
- rejectwater from separators
- aired primary settling tanks
- trickling filters
- sludge thickeners
- anaerobic digesters
- sludge cake hoppers.

4. ABATEMENT OF ODOURS

One of the most important considerations in odour control is the prevention of odour formation, in other words, minimize the potential for septicity, generally by maintaining aerobic conditons. But of course odour problems will rise and, fortunately, today there exist many different techniques to reduce the odour emissions. Some of them are:

- Combustion
- Chemical scrubbers
- Biological scrubbers
- Soil bed filters
- Activated carbon filters
- Iron oxide filters
- Oxidation with chemicals like Ozone, H_2O_2, Chlorine gas.

Combustion is without doubt the best reduction method, but in many instances also the most expensive. If the temperature and contact time of the odorous gases in the combustion chamber are sufficient, the odour compounds would be totally destroyed. Temperatures about 850^0 C and contact time up to 3 seconds have been reported (13,14).

One big problem with combustion of odorous gases is the fuel costs. A special incinerator just to take care of odours from a treatment plant would today not be economical comparable to the use of other reduction techniques. If, however, sludge gases from a digester is available the fuel costs can be cut down and combustion of odour becomes more interesting.

Chemical scrubbers has, perhaps particularly in Norway, became very popular to use at sewage plants. Today about 15 scrubbers are installed. The larger scrubbers (appr. 70.000 m^3/h) clean parts of the total ventilation air at some plants, while the smallest ones (appr. 1500 m^3/h) clean the air from local exhaust systems.

All our chemical scrubbers utilize sodium hypochloride as oxidizing agent. The scrubbers are either single- or two/three-stage scrubbers, and generally it must be concluded that the installation and use of the chemical scrubbers have been quite successful.

One problem with the chemical scrubbers is that they often emit air free from sewage odour but containing "scrubber odour". This "scrubber odour" smells like "chemicals, faint chlorine", and this odour could turn to "typical chlorine" if the scrubbers are uncorrect operated.

Biological scrubbers have now been in work for some years. Instead of washing the odorous gases with chemicals the scrubber tower is filled with activated sludge and then the micro-organisms take care of the odour components. Both an acidic and an alkaline stage are recommended. The odour reduction efficiencies have been reported to be very high, and some chemical scrubbers have been rebuilt to biological scrubbers (15).

Soil bed filters have become more and more popular to use in connection with the reduction of odorous gases. In Germany, for example, large amounts of gases (appr. 100.000 m^3/h) have been cleaned with few problems (15).

When using the filters care must be taken to have the correct soil temperature and moisture content. Very often the soil becomes too dry with bad odour reductions as a result. This is managed by installing water sprinklers that get into function when the moisture drops below a certain value.

Use of carbon filters in odour reduction equipment have several great disadvantages. One is that the odour compounds are not destroyed in the filter, but only restrained until the carbon becomes saturated. Then the possibility occurs that the odour compounds would leave the filter and be a part of the "cleaned" air. Another problem, closely related to the first problem, is to know when the filter is saturated and have to be changed or regenerated. This is impossible to decide without more extensive analyses. The result that can be observed at different plants is therefore carbon filters that either should have been changed or filters that are changed too often; a very expensive process.

Filters filled with iron oxide have been tried with success at some treatment plants in Norway. Wood chips and iron oxide were mixed, air sucked from a sludge tank and finely dispersed in the bottom of the filter box over the filter area. The odour reduction was very good and the filter functioned very well for almost two years without changing of the iron oxide.

Together with the more common use of oxidizing chemicals like ozone, H_2O_2, chlorine gas, the injection of pure oxygen gas into raw sludge has been introduced during the last years. It looks like the technique functions very well, and, as a principle, to avoid odours by maintaining aerobic conditions oxygen addition is an interesting way to go.

5. CONCLUSIONS

Since treatment of wastewater and sludge will nearly always be accompanied by malodours it is of great importance to act as fast as possible when the odour problems arise. One knows today that odour reduction equipment exists and that most of the methods reduce offensive odours more or less.

When deciding what to do with the probmems, the following procedure can be recommended:

1. Look at the treatment plant and the treatment process. Minimize the potential for septicity and maintain aerobic conditions.

2. Determine the exact odour sources, measure the odour strength, and perhaps also analyse the odorous compounds.

3. To clean the total ventilation air can be extremely expensive. Try to use local exhaust systems and eventually clean this decreased amount of air.

4. The results from 2 and 3 are used to decide if odour reduction must take place and also which equipment is best suited.

5. The equipment is installed and the odour reduction efficiencies are measured.

These five remarks have been shown to be very important. It is not enough to trust odour reduction efficiencies given by a manufacturer or odour measurement data from one treatment plant. Two plants very seldom work in the same way, and equipment functioning with extremely good efficiency at one plant does not emply that the same type of equipment functions just as good at another plant.

REFERENCES

1. Rose, L.L.: Odour Control by Chlorination. Sew. Works J. 3 (1931) 462.
2. Dravnieks, A., W.H. Prokop: Source Emission odour Measurement by a Dynamic Forced- Choice Triangle Olfactometer. Air Poll. Control Assoc. Paper (1973) 73-276.
3. Odour Control - a concise guide. Warren Spring Laboratory, England, 1980.
4. Mannebeck, H.: Tragbare Olfaktometer in Geruchsprobleme bei Tierhaltung und Tierkörperbeseitigung. VDI-Berichte 226, VDI-Verlag, Düsseldorf, 1975, 103.
5. Compilation of Odour and Taste Threshold Values Data. ASTM Data Series DS48A, 1978.
6. Compilation of Odour Threshold Values in Air and Water. National Institute for Water Supply and Central Institute for Nutrition and Food Research TNO, Netherlands, 1977.
7. Odors emitted from raw and digested Sewage Sludge. EPA-670/2-73-098, Washington, 1973.
8. Ando, S.: Odor Control of Wastewater Treatment Plants, J. Water Poll. Control Fed. 52 (1980) 906.
9. Henny, J.G. and R. Gehr: Odor Control: an Operator's Guide. J. Water Poll. Control Fed. 52 (1980) 2523.
10. Hangartner, M.: Sampling and Analysis of Odorous Compounds in Water Treatment Plants. Inter. J. Environ. Anal. Chem. 6 (1979) 161.
11. Hangartner, M. and H.U. Wanner: Geruchsanalysen bei Abwasserreinigungsanlagen. Wasser, Energie, Luft 70 (1978) 3.
12. König, W.A. et al.: Schwefelverbindungen - Verursacher von Betonkorrosion? EWPCA-State-of-Art-Seminar, Corrosion in Sewage Plants, Hamburg, 1982. EWPCA '82 Documentation, 13.
13. Pettit, C.G.: 20 Years of Sewage Sludge Burning at Barberton, Ohio. J. San. Eng. Div. Amer. Soc. Civil Engr. 85SA6 (1959) 17.

14. Laboon, J.F.: Construction and Operation of the Pittsburgh Project. J. Water. Poll. Control Fed. 33 (1961) 758.
15. Pfeiffer, H.: Lecture at the Colloquium "Minderung von Geruchsstoffemissionen aus Stationären Anlagen", Wiesbaden, 1981.

DISCUSSION

H. Scheltinga

When you refer to soil bed filters, I think you do not mean normal soil but town compost, as used in Germany and Holland, which has proven very effective and economical.

N. Berg

What I mean by soil is normal natural soil, which works very well for odour removal. It has also been used in the United States. It requires higher air-compression and of course there may be problems with blowers and corrosion, which I have not discussed in detail.

A.M. Bruce

In the United Kingdom, we have also found soil filtration useful for odour removal. You indicated that H_2S may not be the main problem, which is surprising. Finally, could you comment on the selection of your odour panel members.

In many cases polysulfides may be a more significant source of odour than H_2S. They produce a more intense odour and are mainly responsible for corrosion. Our odour panel members, who are volunteers and number about 15, are selected from our own staff. They are trained, using standard substances such as pyridine. We have found that motivation is the most important factor.

SESSION II

Chairman: Prof. R. LESCHBER

Rapporteurs: Prof. T.J. CASEY
Dr. P.J. NEWMAN

CHARACTERIZATION OF THE PHYSICAL STATE OF SLUDGES

F. COLIN
Institut de Recherches Hydrologiques
Nancy, France

Summary

The variation of the physical properties of sludges according to their concentration is so large that it is necessary to define several classes of physical condition, for each of them it is possible to apply homogenous characterization methods.

After a literature review of the various flow schemes, one has studied the application of sludge rheology to the study of their structure and treatability.

Then is defined the application of rheologic characterization to the dimensioning of the devices for sludge transfer in pipelines, and the conception and the use of viscosimeters destined to the study of waste sludges.

The characterization of sludges not susceptible to show a continuous fl is then briefly studied. A synthesis exposes the relationships between the characteristics of sludge physical condition and the selection of their mode of transport, stocking and application.

1. INTRODUCTION

The physical state of a sludge and particularly its mechanical properties are of considerable importance since they govern its handling, transport, storage and disposal.

A knowledge of the analytical characteristics and the water content of a sludge is quite inadequate for predicting its physical and mechanical methods and tests.

In practice the techniques used to characterize the physical state of sludges can be applied with a view to :

a. obtaining a better knowledge of the current state of a sludge ;
b. studying the structure of a sludge, ie the interactions between the constituent solid particles ;
c. predicting the behaviour of the sludge during handling and transport and, if necessary, designing the equipment required.

After defining the chief categories of physical states of sludges, we shall seek to ascertain valid methods of characterization within each of these categories and the possibilities of using the results for studying and predicting the behaviour of sludges.

II. DEFINITION OF THE CATEGORIES OF PHYSICAL STATE OF SLUDGES

It is conventional to categorize sludges into the following four physical states (2) :

1. Liquid state Sludge which flows under the effect of gravity. Flow is the best method of handling.
2. Plastic state The state in which the material may undergo permanent deformation if the mechanical stress applied to it exceeds a certain treshold. Sludges in this state are capable of continuous flow if sufficient pressure is applied.
3. Solid state with shrinkage (sometimes incorrectly called the friable solid state). This is a solid state in which the volume decreases as drying progresses, ie it relates to a medium still saturated with water.
4. Solid state without shrinkage Below the moisture content level characterizing the limit of the preceding state, the volume of the sludge no longer decreases as drying progresses, ie the medium is no longer saturated with water. The sludge is then in the solid state without shrinkage.

The work of the Institut de Recherches Hydrologiques (Hydrological Research Institute) (2) has made it possible to develop simple tests for fixing the limits of the conventional physical states and selection criteria by means of which a given sludge can be placed into one of the four categories defined above. The tests, or a combination of them, give "yes or no" answers as to whether or not a given sludge corresponds to one of the physical states in question.

Test defining whether a sludge corresponds to the liquid state : a test has been developed which results directly from the definition of the liquid state, being a FLOW TEST consisting in determining whether the sludge can flow through a calibrated orifice upstream of which the sludge is subjected to a given load corresponding to the weight of the column which it forms. Conventionally the sludge is referred to as liquid if all of it flows up to the level of the top of the orifice within a given time.

Test defining whether a sludge corresponds to the plastic state : by convention a sludge is said to be plastic if simultaneously it gives :

a. a negative result to the preceding test (no flow) ;
b. a positive result to a PLASTICITY TEST, ie has a moisture content such that it can be formed into small spindle-shaped pieces without these crumbling any further.

Another method consists in replacing this purely manual test by an EXTRUSION TEST in which pressure is applied to the sludge by a plug so as to extrude it through a calibrated tube 3 mm in diameter and 20 mm long. In this case the sludge will be said to be in the plastic state when it can be extruded in the form of a constinuous cylinder 3 mm in diameter when sufficient pressure is applied.

Test defining whether a sludge corresponds to the solid state with or without shrinkage : a sludge corresponds to one of these conventional physical states when it gives negative results to the two above-mentioned tests (neither liquid nor plastic). The criterion for deciding between the two solid states is provided by a shrinkage test consisting in observing any reduction in volume of the sludge during drying.

Although the diverse nature of the mechanical properties of sludges corresponding to different physical states makes it impossible to compare them with each other owing to the lack of common methods of characterization, such a comparison can be made within each physical state or for groups of sludges exhibiting homogeneous behaviour.

In the following we shall distinguish:

a. the characterization of sludges capable of continuous flow, whether they be liquid or plastic ;
b. the characterization of non-liquid sludges, whether plastic or solid.

III. CHARACTERIZATION OF SLUDGES CAPABLE OF CONTINUOUS FLOW

Liquid and plastic sludges capable of continuous flow under the effect of gravity or of an applied pressure are best characterized using rheological methods, and more particularly by the study of flow properties.

As early as 1934, Merkel (3) began studying the behaviour of Imhoff sewage sludges with a rotating viscometer, demonstrating the phenomenon of thixotropy and the fact that an increase in the hydrophilic particle concentration in the sludge increased the viscosity to a far greater extent than did an increase in the hydrophobic particle concentration.

In 1936, W.D. Hatfield (13) confirmed the phenomenon of thixotropy (decrease in the viscosity of the sludge following previous agitation) and the relationship between viscosity and concentration. He proposed a pseudo-plastic model and the relationship between viscosity and shear rate as represented by the equation :

$\eta = a\,\gamma^n$, where η = viscosity
γ = shear rate
n = -1.1 to -1.4
a = a constant dependent on the solids concentration and its organic fraction

In 1950, Wolfs (14) studied the factors affecting the flow of sludge in force mains. He observed a considerable head loss at low velocities and a small head loss at high velocities ; he also noted that this latter head loss was not much greater than that which occurred with water.

Then in 1958 Tsung-Lien (15) attempted to define a limiting water content for sewage sludge below which the flow was plastic and above which the material was similar to a Newtonian fluid. He supported this principle with the argument that in very low-concentration sludges the solid particles do not interact with each other, the sludge behaving like a free suspension. "The presence of particles in suspension can thus only increase the density of the fluid and has only a negligible effect on viscosity".

In 1962, Behn (5, 6, 16) compared cylinder viscometers and capillary viscometers for determining the flow properties of sludge, and opted for capillary viscometers (or more accurately tube viscometers, the diameter of the tubes being such that they cannot properly be called capillaries). These viscometers make it possible, by varying the length of tubes of the same diameter, to demonstrate any thixotropic phenomena.

On the basis of the results obtained, Behn developed a mathematical model and found a more general form for the equation known as the "power law" :

$$\tau = 10^{bo} C^{b1} (\gamma)^{b2 + b3 \log C}$$

where C = concentration (SS percentage)
τ = shear stress (dynes/cm^2)
γ = shear rate (sec^{-1})

A large number of authors subsequently studied the flow laws of sludges and characterized their rheological behaviour in this way.

Table 1 (30) summarizes various types of rheological behaviour demonstrated for sewage sludges (mainly urban sludges) while Table 2 (3) presents a general classification of rheological models, distinguishing those cases in which the flow limit τ_o is zero (the fluid becomes deformed as soon as a mechanical shear stress is applied) and those in which τ_o is not zero (it is necessary to apply this minimum stress before the fluid begins to deform).

IV. APPLICATIONS OF THE RHEOLOGICAL CHARACTERIZATION OF SLUDGES CAPABLE OF FLOW

The following applications have been examined by different authors :

a. study of the structure of sludges, the size and the cohesion of flocs of agglomerated particles in suspension ;
b. study of the treatability of sludges by seeking to establish relationships between the rheological characteristics of sludges and features of treatability (suitability for thickening, flocculation, centrifugation, etc.);
c. study of the suitability of sludges for pumping and for transport in pipes.

We shall now examine these various categories of application in turn.

IV.1 Study of the structure of sludges

When suspensions of solid particles exhibit Newtonian behaviour their viscosity (τ/γ ratio) increases with the concentration of suspended matter. In fact, it is not the concentration per unit mass but the concentration per unit volume which is relevant (ϕ).

According to Einstein, in the case of a dilute suspension of spherical particles without any mutual interaction :

$$\eta = \eta_o (1 + 2.5 \phi)$$

η_o being the viscosity of the dispersion liquid.

TABLE 1 - SUMMARY OF THE TYPES OF RHEOLOGICAL BEHAVIOUR FOUND FOR SEWAGE SLUDGES BY DIFFERENT AUTHORS

Author	Type of behaviour	General equation	Values of constants found
Hatfield (13)	Pseudo-plastic Ostwald	$\eta = a\gamma^n$	$n = -1.1$ to -1.4 a = not specified
Behn (5)	Pseudo-plastic Ostwald	$\tau = k\gamma^n$	$n = 0.8169 - 0.8426 \log C$, $K = 0.0113C^{3.3954}$
Geinopolos and Katz (42)		$\eta 1_t = K_T\ C^{3,3}$	K_T = not specified
Dick and Ewing (7)	Plastic Bingham	$\tau = \tau_o + \eta\gamma$	$\tau_o = J\ e^{KC}$ j,k = not specified
Colin (1)	Plastic Casson	$\sqrt{\tau} = \sqrt{\tau_o} + \sqrt{\eta}\ \sqrt{\gamma}$	$\tau_B = KC^n$ $n = 3.75$ K = not specified
Bhattacharya (38)	Pseudo-plastic Herschel-Bulkley No 1	$\tau = \tau_o + k.\gamma^n$ (for primary sludge)	$K = 0.008$ to 0.113 $n = 0.78$ to 0.94 depending on the concentration of the sludge
Chavarria (30	Pseudo-plastic	$\text{Log}\ \gamma = a_o + n\ \text{Log}\ \eta$	$n = 0.058\ C - 1.642$ $a_o = b + d \log C$ $b = 0.494 + 0.044\ (\%\ VM)$ $d = 41.06 - 1.0223\ (\%\ VM) + 0.0068\ (\%\ VM)$

In all these relationships, C = SS percentage
VM = volatile matter percentage

TABLE 2 -

	Authors	Rheological equation
Models without flow limit ($\tau_o = 0$)	Newton	$\tau = \mu \gamma$
	Ostwald - de Waele	$\tau = k\gamma^n$
	Ellis - de Haven	$\tau = \frac{\mu_o}{1 + C \tau^n} . \gamma$
	Reiner- Philippoff	$\tau = (\mu_\infty + \frac{\mu_o - \mu_\infty}{1 + (\frac{\tau}{A})^2} . \gamma)$
	Sisko	$\tau = A \gamma + B \gamma^n$
	Prandtl - Eyring	$\tau = A \sin h^{-1} (\frac{1}{\beta} \gamma)$
	Powell - Eyring	$\tau = C\gamma + D \sin h^{-1} (\frac{1}{E} . \gamma)$
Models with flow limit ($\tau_o \neq 0$)	Bingham	$\tau = \tau_o + \eta \gamma$
	Herschel-Bulkley No 1	$\tau = \tau_o + (\eta \gamma)^{1/m}$
	Herschel-Bulkley No 2	$\tau = \tau_o + \frac{\eta_o}{1 + C(\tau - \tau_o)^n} . \gamma$

In the case of non-spherical particles, the value 2.5 has to be replaced by a form factor k.

For more concentrated suspensions, various authors have formulated empirical relationships of the general type (10) :

$$\eta = \eta_o \, [1 + k_1 \, \emptyset + k_2 \, \emptyset^2 + A \exp. \, (B \, \emptyset)]$$

the constants k_1, k_2, A and B varying according to author.

The work of Michaels and Bolger (11) and then of Hunter and Nicol (12) carried out on kaolinite sludges demonstrated the effect of the surface electric charge of the particles on the flow properties of this type of sludge, and can be generalized to cover the case of suspensions having a tendency to rapid flocculation. In this case the rheological properties are closely related to the intensity of the interactions (attractions) between the solid particles in suspension.

This work was used as a basis by Colin (1) to study the structure of flocculated sewage sludges from rheological curves plotted using a tube viscometer.

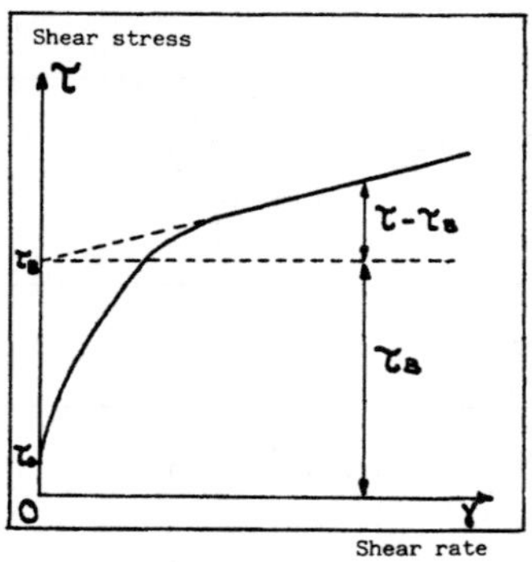

Figure 1 : Example of sewage sludge rheogram

Such a curve (rheogram) may be interpreted as follows :

a. The behaviour of the sludge under study is not Newtonian since the rheogram is not reduced to a straight line.
b. The extension of the curve on the co-ordinate-axis origin side does not pass through zero but intersects the shear-stress axis at a point τ_o. A minimum shear stress therefore needs to be exerted in order to initiate deformation of the fluid. τ_o is called the flow threshold or limit : its existence has to be related to the particular structure of the sludge, which consists of aggregates of elementary particles, and these aggregates are sufficiently close together to form a continuous three-dimensional network which has to be broken in order for flow to occur.
c. The apparent viscosity or slope of the $\tau = f(\gamma)$ curve is not constant, and decreases gradually with increasing shear stresses. This variation indicates of a structural modification of the suspension under the effect of the stresses exerted upon it : there is a decrease in the concentration per unit volume of flocculent particles which break up and disperse in the free water.

The completion of aggregate breakdown will be indicated by a constant apparent viscosity corresponding to the final linear section of the $\tau = f(\gamma)$ curve. This "Bingham" viscosity is given the notation η_B and the extension of the linear section of the rheogram intersects the shear-stress axis at the point τ_B.

The stress τ_B , starting from which the rheogram becomes linear, constitutes the minimum shear stress that needs to be applied to the fluid in order to ensure that the individual particles are totally dispersed in the suspension.

Ultimately, the intensity of the forces of attraction between particles in the chief factor affecting the relationship between the shear stress τ and the deformation rate γ, and rheological curves make it possible to define a specific quantity characterizing the mechanical behaviour of sewage sludges. The importance of this work will be immediately realized when it is pointed out that this property is one of the chief factors affecting their suitability for dewatering.

Colin (1) showed that τ_B varied with the concentration C of the suspension in accordance with a law of the following form :

$\tau_B = KC^n$, where n = 3.75 for the sludge in question.

In these conditions the constant K is a measure of the force of attraction between the particles of the suspension.

Various methods have been developed for measuring floc solidity or the mechanical shear strength of flocculated sludges :

a. work by Gale and Baskerville in 1972 (measurement of the capillary suction time after agitation in clearly defined conditions as regards duration and intensity).
b. work by Colin (19) (changes in the mechanical power dissipated as a function of time in the case of agitation of uniformly constant intensity in a suitable reactor) ;
c. more recent work by Campbell dand Crescuolo (24) (rheogram plots for various chemical conditioning treatments of an identical anaerobically digested sludge, and study of the effect on the rheograms of the operational parameters of conditioning : duration and intensity of stirring, dosage and type of reagents).

Lastly, in special fields such as oil-drilliing sludges, attempts have been made to determine the relationship between the rheology of these sludges and their composition (45).

IV.2 Study of the treatability of sludges

A number of attempts have been made to establish relationships between the rheological characteristics of sludges and their suitability for various treatments.

IV.2.1. Relationships between rheology and settleability or susitability for thickening

In 1967, Dick and Ewing (7) studied the rheology of activated sludges using rotating viscometers with coaxial cylinders, with a view to relating this aspect to their studies on thickening. These sludges, at the concentrations existing in aerotion tanks, exhibit plastic and even thixotropic behaviour. They found that the rheology of these sludges provided an explanation for the divergence dbetween their behaviour and that which could be predicted using conventional gravity-thickening models.

Tests carried out at the Water Research Centre (46) indicated a bilogarithmic relationship between the viscosity of an activated sludge at a velocity gradient of 100 s^{-1} and the settleability index. This relationship is represented in graph form below :

RELATIONSHIP BETWEEN THE SETTLEABILITY OF ACTIVATED SLUDGE TREATING SETTLED SEWAGE, AND THE VISCOSITY OF THE SLUDGE AT A SUSPENDED SOLIDS CONCENTRATION OF 3 PER CENT.

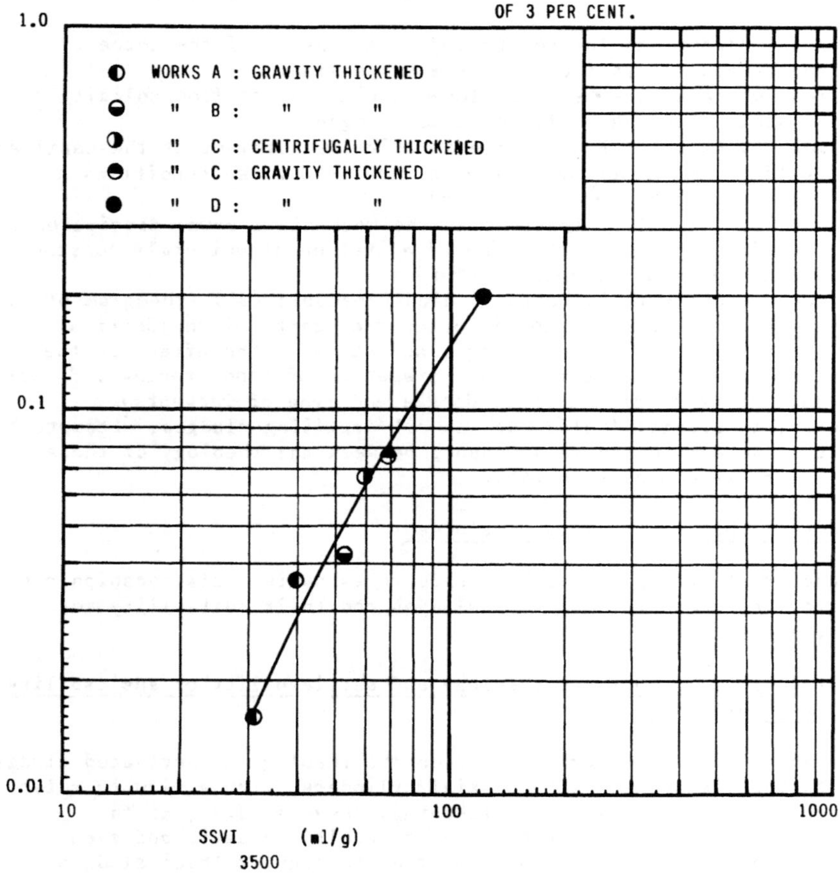

Flocculation of activated sludges by controlled and optimized agitation between the activation tank and the clarifier of a treatment plant facilitates the settleability of these sludges (40).

Chavarria (30) was able to demonstrate relationships between the thixotropic properties of activated sludges and the structural characteristics of the floc (photomicrographs) and to give values for capillary suction times. He did not, however, establish any relationship between the sedimentation properties of sludge and its viscosity.

IV.2.2. Relationship between the rheology of sludges and their floatability

In 1964, Geinopolos and Katz (47) sought to establish a relationship between the rheology of activated sludges and the performance of a sludge collector using the dissolved-air flotation process (rotating cylinder system). This enabled them to demonstrate the bilogarithmic relationship represented below :

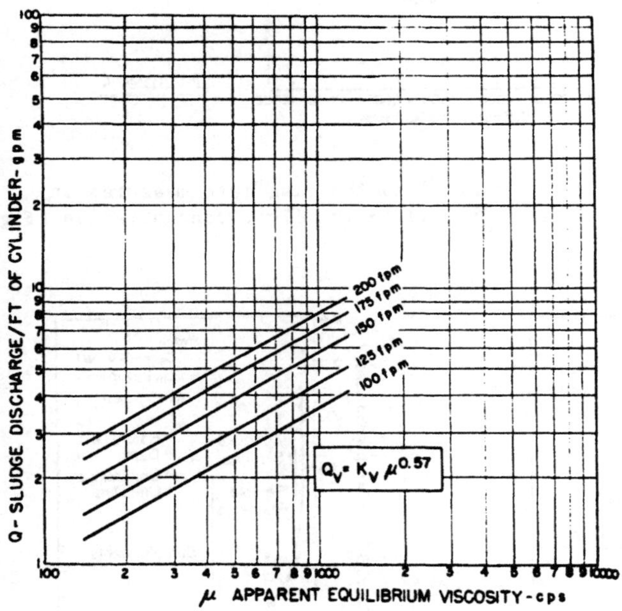

Figure 3

More recently, in 1975, Wood and Dick (35) looked into the use of air-dissolved flotation for the thickening of surplus activated sludges, mainly in the case of sludges exhibiting the phenomenon of bulking. They were able to show that the sludges in question exhibited a plastic Bingham behaviour. The flow limit values varied as a function of the concentration of the sludge in accordance with a relationship depending on the nature and morphology of the micro-organisms involved (Figure 4).

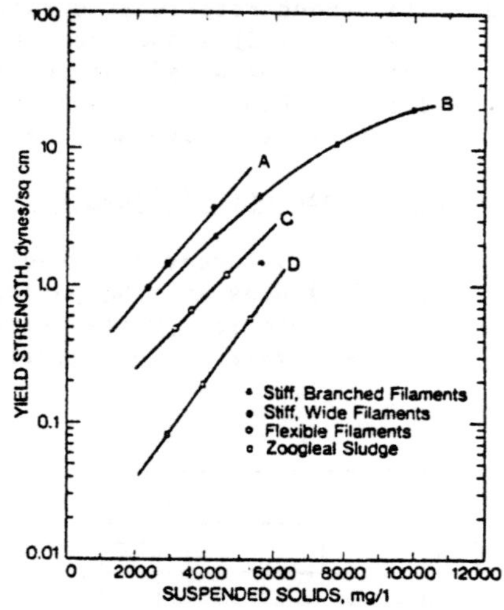

Figure 4

This flow limit is itself related to the rise rate measured in discontinuous processess (Figure 5) and to the float concentration (Figure 6) in constinuous procésses.

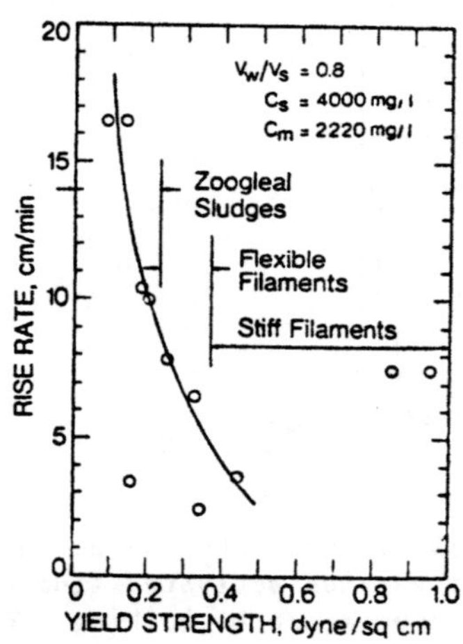

Figure 5

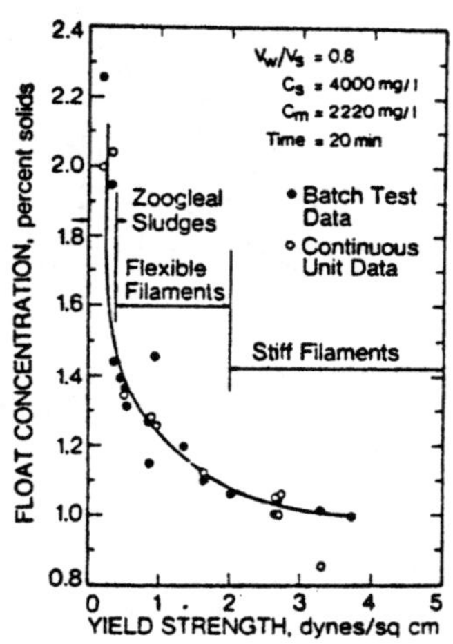

Figure 6

IV.2.3. Relationships between rheology and anaerobic digestion

Numerous general studies not specifically related to sewage sludge treatment have been carried out on the rheology of media during fermentation (see reference (32) for instance).

The reason is that these rheological properties govern the agitation and mixing possibilities of the medium and also mass exchanges and transfers between the solid, liquid and gaseous phases.

In the more specific field of the anaerobic digestion of sludges, Colin (43) looked into the maximum permissible solids concentration for treating industrial sludges with a view to increasing energy production by working with a very high volume load.

Generally speaking, a fairly abrupt transition occurs between the liquid state and the sticky state, as a function of the solid particle concentration of sludges. The concentration of water content value corresponding to this transition depends to a large extent on the nature of the substrates (concentrations of 15 % and 21 % for the two substrates studied). In these conditions the energy required for maintaining constant agitation increases rapidly with the concentration in suspension.

For example, it is found that, in the case of a substrate before fermentation, increases in concentration from 5 to 9 %, then from 9 to 11 % and finally from 11 to 12 % require increases in power from 5 to 10 W/m^3, from 10 to 30 W/m^3 and from 30 to 95 W/m^3 respectively.

The disappearance of solid substrate during the acidogenic phase produces a radical change in the rheological characteristics of the medium as it becomes fluid. These changes are more pronounced than could be expected merely as a result of the reduction in concentration of the solid phase, which constitutes proof of a simultaneous change in the structure of the suspension and, more particularly, the physical interactions between its constituent particles.

The flow limits τ_0 may fall by a factor of 20, or may even become zero, while the power required for maintaining identical agitation becomes considerably less. An example is given in Figure 7, this agitation power being expressed as a function of the solids concentration of the sludge for different stirring intensities and for the same sludge before methane fermentation (curves A1, B1, C1, D1) and after it (curves A2, B2, C2, D2).

IV.3 Application to the design and dimensioning of equipment for pumping and conveying sludges through pipes.

A literature survey on sewage sludge pumping was made by Willis (34). As he states, many authors point out the effect of the rheological or physical characteristics of sludges on pump performance (48 to 53), but none of them gives any precise quantitative information. It seems that for solids concentrations of less than 2 or 4 %, depending on the author, sludges behave like water (Newtonian fluid), whereas above 10 % it is necessary to use volumetric pumps (single- or double-acting piston or helicoidal rotor) or rotodynamic pumps with radial, axial or mixed flow. Highly concentrated sludges (mechanically dewatered, with a concentration of 20-45 %) can be transported by eccentric helicoidal rotor pumps (Moineau type) provided with a hopper and fed by a screw-drive.

Selection criteria for pumps based on various considerations, especially rheological ones, were defined by Frost (26, 31). The chief criteria are :

a. the rheological characteristics ;
b. the sludge concentration variability ;
c. the hydrodynamic range chosen (laminar or turbulent) ;
d. the gas concentration and the risks of outgassing ;
e. the presence of fibres or other clogging materials.

The Water Research Centre (26) recently developed a simple method for evaluating the rheological properties of sludges from tests using a laboratory centrifuge designed for low-acceleration thickening studies. The evaluation can also be made from direct rheological measurements.

The behaviour of a sludge can be assessed by the head loss ratio (HLR), which is the ratio of the head losses measured with the sludge in question and with water in the same pipes carrying these two fluids at the same velocity.

The surest method of choosing pipe transport dimensions consists in evaluating the head losses in a pipe loop for different flow rates. The error which consists in calculating pipe configurations by simply taking for the head loss a multiple (eg x 6) of the head loss obtained with water increases with decreasing flow velocity and with laminar flow (27). The HLR drops in the turbulent flow mode. Tests carried out for the North West Authority by the Binnie company in the UK and the work of the Water Research Centre (27) have shown the following :

a. the most suitable rheological model of sludges for representing experimental results and dimensioning pumping and transport systems is the Bingham model generalized by Herschel-Bulkley ;
b. for the laminar mode, results obtained with tube viscometers can be extrapolated to the calculation of full-size pipes ;
c. after the application of mechanical shear stresses, sludges do not regain any appreciable viscosity for a period corresponding to normal non-pumping periods in full-size installations ;
d. sludges having a higher than average concentration of biological or fibrous matter exhibit a high resistance to flow in the laminar mode ;
e. mixtures of sludges have rheological properties that are intermediate between those of the constituent sludges ;
f. when the flow passes from the laminar mode to the transition mode the HLR drops until it reaches a value of the order of 1.3. ; it then tends towards 1 when the flow becomes strongly turbulent ;
g. for a given viscosity the power losses of rotodynamic pumps are greater for primary sludges than for anaerobically digested sludges, and for digested sludges than for activated sludges ;
h. the following empirical relationships were established in tests by the Water Research Centre at Stevenage :
    i. digested sludges with a concentration between 1.8 and 6.2 %: HLR = 0.8 + 0.16 (concentration in percentage) ;
    ii. activated sludges with a concentration between 0.5 and 4 %: HLR = 0.88 + 0.24 (concentration in percentage) ;
    iii. primary sludges with a concentration between 2.5 and 5.8 % : HLR = 1.0.

As regards the calculation of head losses through pipes on the basis of rheological characteristics, a survey has been undertaken by Willis (34), who drew a distinction between theoretical studies, studies on actual sites and laboratory experiments. Rational methods for operational purposes have been defined by the Water Research Centre (29) on the basis of viscosity-measuring tests carried out with a tube viscometer (diameter 12/27 mm) and with a pipe 107 mm in diameter.

For the laminar mode, viscosity-measuring tests make it possible to assess full-size head losses to within +/- 20 %, whereas for the turbulent mode the best solution consists in evaluating head losses by taking those which would be calculated for water and multiplying them by an empirical factor selected as a function of the type of sludge. Simple criteria make it possible to determine whether the flow will be laminar or turbulent.

Work of the same type has been carried out in the USA by Mulbarger et al (25), who made a survey of literature published over the last 50 years on the use of rheology in calculating head losses for the transport of urban sewage sludges. In addition to confirming what has been stated above these authors also showed that in the laminar mode the head loss was independent of the type of pipe material, contrary to the case of the turbulent mode.

Finally, in the field of chemical engineering, we find such work as that carried out by Cheng (36) on the design and dimensioning of pipes intended for conveying non-Newtonian fluids.

V. DESIGN AND USE OF VISCOMETERS FOR STUDYING SEWAGE SLUDGES

Viscometers which can be used for studying sewage sludges fall into the following two general categories :

1. rotating-type ;
2. tube-type (incorrectly referred to as capillary viscometers).

They are used to determine the flow properties of sludges and thus to deduce their rheological properties. Behn (6) indicated the flow relationships and laws of rotating viscometers with coaxial cylinders and of tube viscometers.

Rotating viscometers may have :

a. coaxial cylinders with rotating inner cylinder ;
b. coaxial cylinders with rotating outer cylinder ;
c. rotating blades.

The drawback of coaxial-cylinder viscometers is that the cylinders have to be very close together when studying sludges which are not very concentrated and/or not very viscous. Consequently there is a risk of clogging by sand grains, fibres and other solid materials. Another drawback is the phenomenon of slip which may occur at the cylinder/liquid interface and the separation of particles in the gravitational field and under the effect of the centrifugal field induced by the rotation of the cylinder. Some authors remedied this by artificially increasing the roughness of the rotating cylinder.

In the case of viscometers with blades, the velocity gradient is less well defined so that all that is obtained is a mean value based on the mechanical energy dissipated in the medium (calculated from measuring the drive torque of the mover). Many rotating viscometers are available commercially and are of varying adequacy for the rheological characterization of sludges.

Tube viscometers are more consistently reliable. The tube diameter must be large enough to prevent any clogging phenomenon, which means that the tube has to be long in order to obtain measurable head losses. These systems are not commercially available and have to be specially constructed. Figure 8 shows a complete assembly such as that used at the Water Research Centre.

There are systems (2, 28) in which a constant sludge feed can be achieved by gravity, and thre are also simplified devices (2, 33) which measure the throughput of sludge flowing through a calibrated orifice or a very short tube fed under load by the sludge.

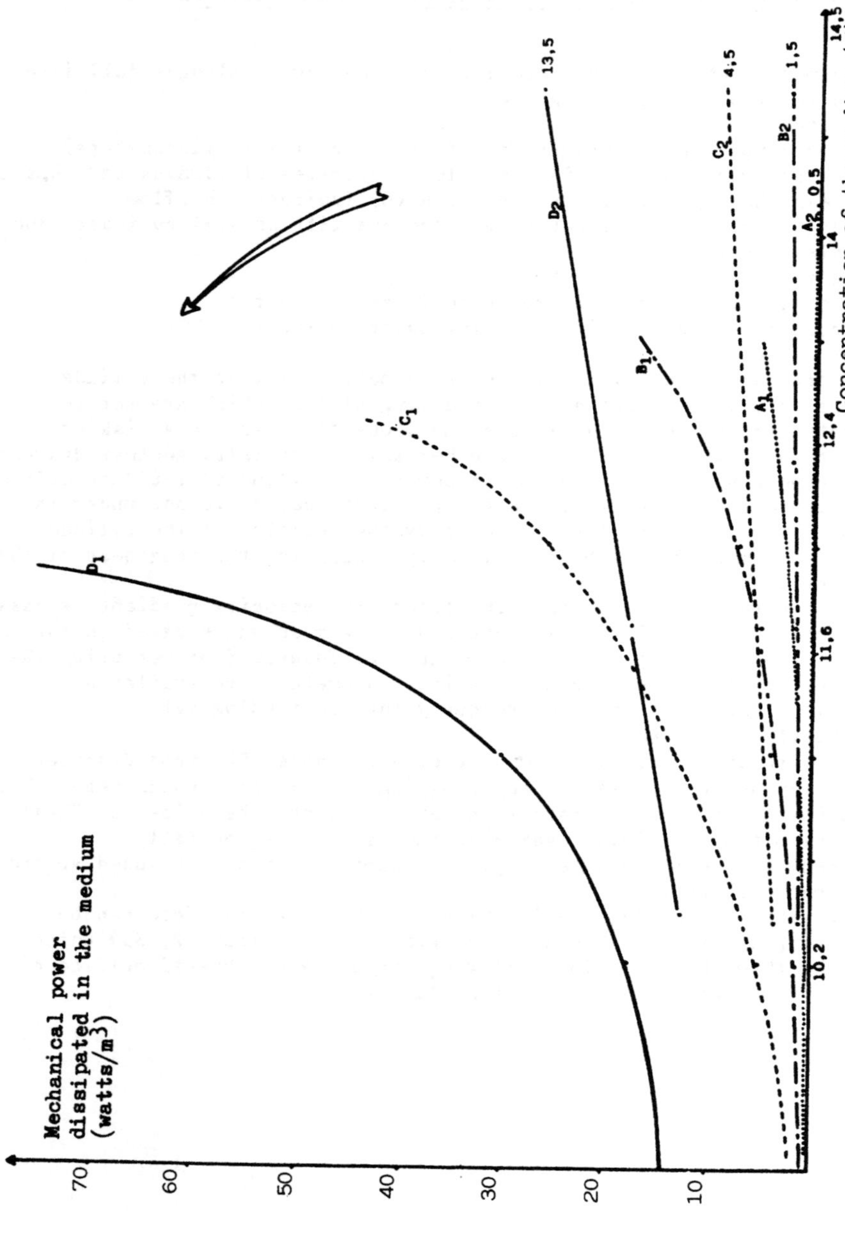
Mechanical power
dissipated in the medium
(watts/m3)
70
60
50
40
30
20
10
D1
C1
D2
B1
A1
C2
B2
A2
13,5
4,5
1,5
0,5
14,5
14
12,4
11,6
10,2
Concentration of the medium (%)

Figure 7

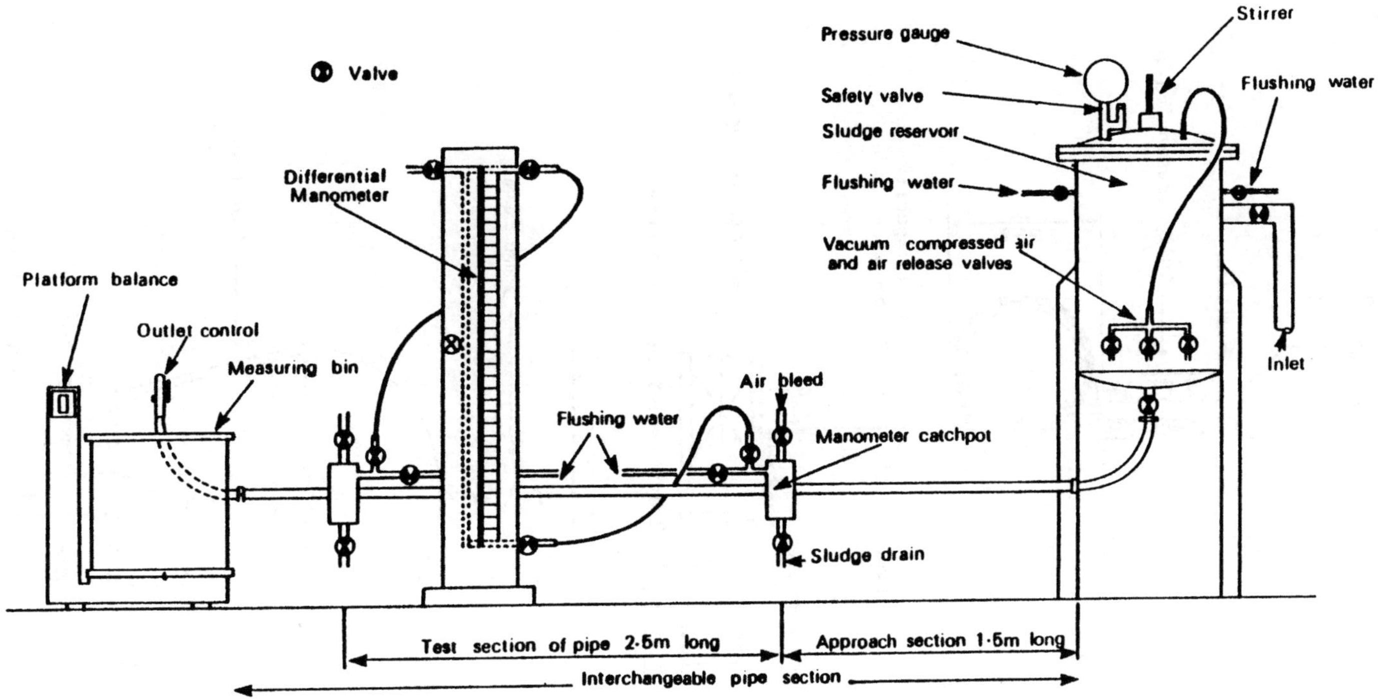

Figure 8 Equipment layout for a tube viscometer

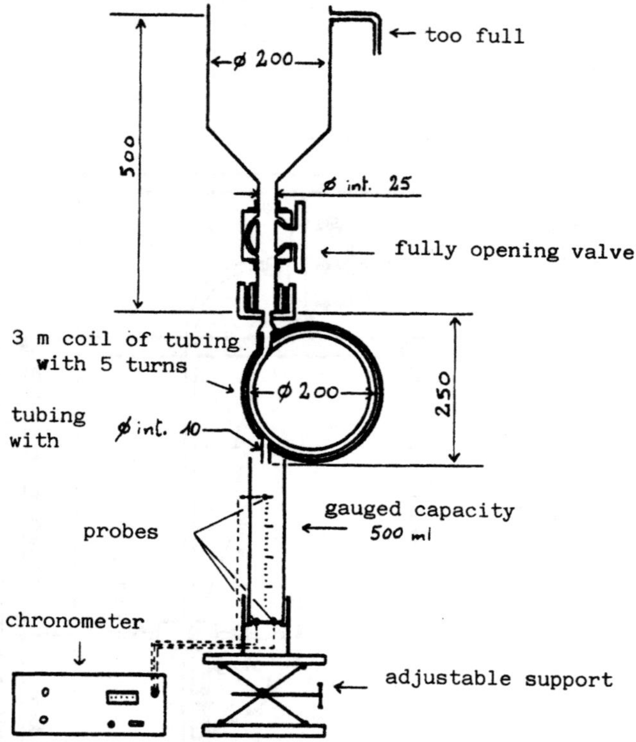

Figure 9 Gravity-feed tube viscometer (2)

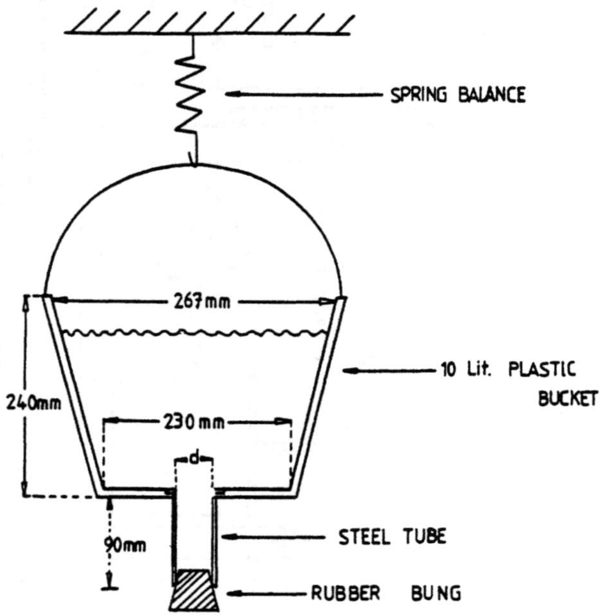

Figure IO Viscometer with calibrated orifice placed under load (33)

Tube viscometers have been successfully adapted to measure the characteristics of slowly settling flocculated suspensions (39).

VI. CHARACTERIZATION OF SLUDGES NOT CAPABLE OF CONTINUOUS FLOW

Here we are concerned with characterizing sludges in the sticky or solid state. A sticky sludge is said to be in the plastic state if it can be formed into small spindle-shaped pieces without these crumbling any further. As this concept is difficult to quantify, Colin (2) replaced this purely manual test with an EXTRUSION TEST using a calibrated tube 20 mm long and 3 mm in diameter through which the sludge is extruded under pressure applied by a piston. In this case the sludge will be said to be in the plastic state if it can be extruded in the form of a continuous cylinder 3 mm in diameter when sufficient pressure is applied (Figure 11).

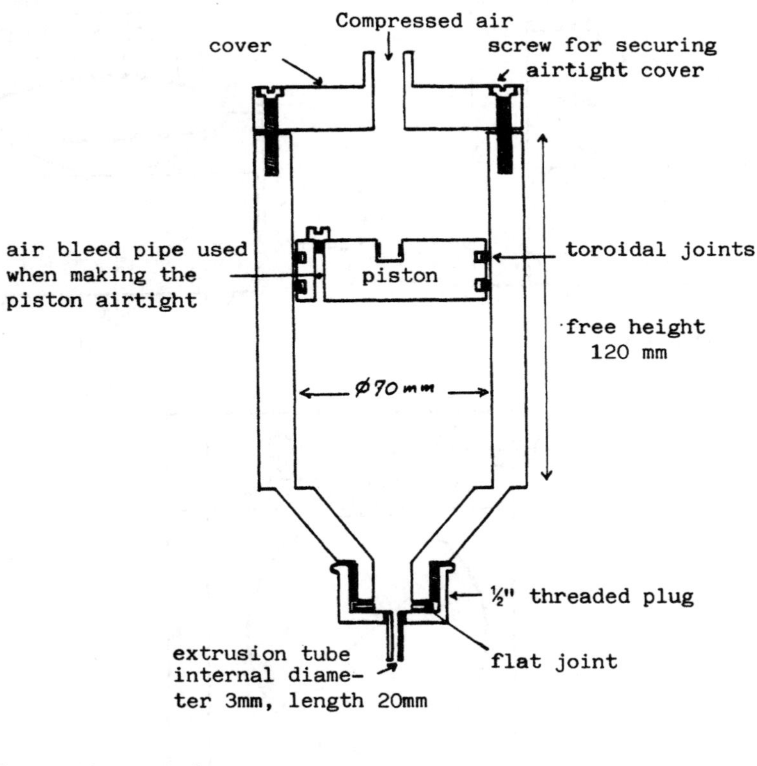

Figure 11 Extrusion test apparatus

The device can also be used for quantitative characterization of the degree of plasticity and consistency of sludge. The test consists in extruding the sludge at a relatively rapid rate by applying a high pressure and then slowly and gradually reducing the pressure until extrusion stops. The pressure value corresponding to the cessation of extrusion provides a measure of the degree of plasticity of the sludge.

The same author (2) developed a MECHANICAL STABILITY TEST.

This is designed to reveal the mechanical behaviour of the sludge when exposed to successive impacts. The Casagrande apparatus used (54) is shown in Figure 12 below. The test consists in determining the number of impacts required to make the two lips of a groove made in the sludge with a standard tool join up over a given length (fixed at 1 cm). The larger the number of impacts required, the greater is the mechanical stability of the sludge.

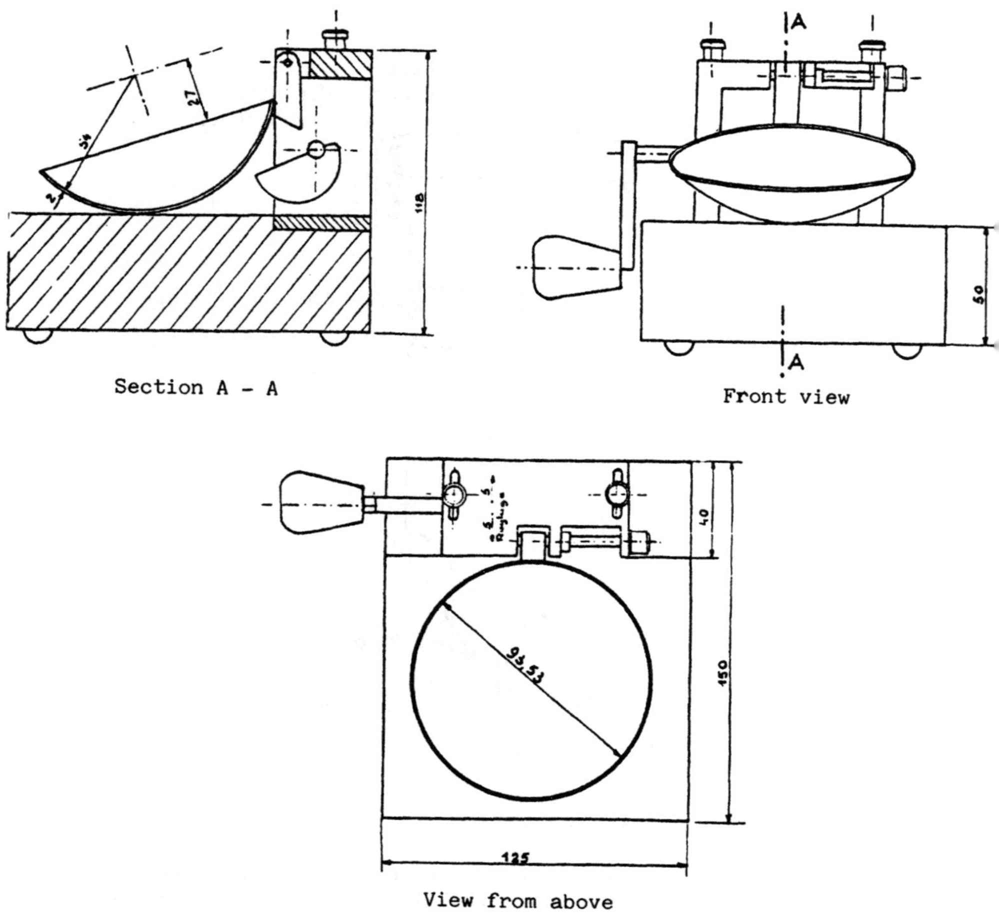

<u>Figure 12</u> Casagrande apparatus

The apparatus consists of a rough-bottomed dish which pivots about a horizontal spindle and to which a cam mounted on a handle applies impacts of a fixed amplitude (10 mm) at the frequency wished by the operator (two blows per second). The dish is completely filled with sludge and the surface is levelled off. Using a standard tool of triangular cross-section (ASTM gauge D 423-39) a diametrical groove of constant width is made in the plane perpendicular to the spindle. The test consists in determining the number of impacts necessary to make the two lips of the groove join up over a given length (fixed at 1 cm).

The larger the number of impacts required, the greater is the mechanical stability of the sludge.

The plastic state is characterized by permanent deformation when sufficient mechanical stress is applied to the material, and consequently the best indication of this state is obtained by measuring the deformation produced by a given stress. A manual penetrometer may be used for this purpose, the measuring needle being replaced by a calibrated cone (2).

The apparatus and the way in which it functions are illustrated below (Figure 13).

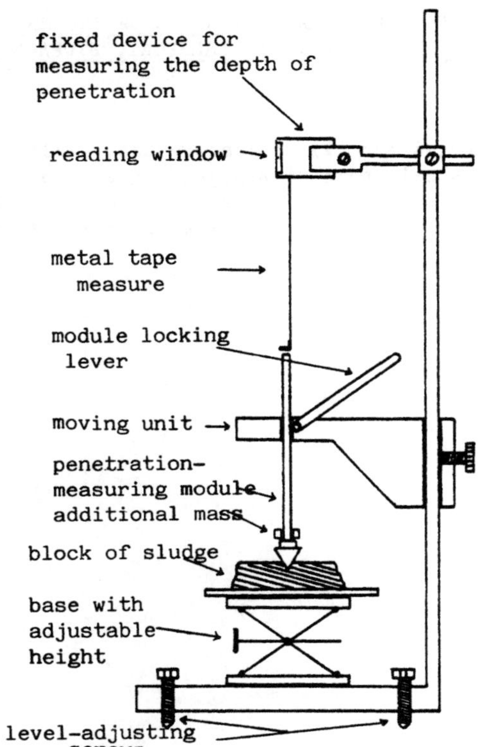

Figure 13 Diagram of penetrometer

The procedure involves measuring the imprint produced by the cone after it has been applied under several different loads. Various devices with different angles of conicity (30°, 45°, 60°) are used to cover the plasticity range encountered with sludges. For solid sludges a cone with a very small angle (needle penetration meter) is required to obtain measurable penetration.

Adhesion is another important practical consideration : the adhesion of sludges to tools or handling equipment can be a serious nuisance. This property of sludge can be assessed by means of a laboratory test (2).

The test is carried out with a cone with an adequate angle of opening, which is fully submerged in the plastic sludge so that the top is flush with the flat surface. The test consists in using a dynamometer to determine the force required to release the cone when it is subjected to a very gradual vertical movement. The weight of the cone itself is deducted from the value obtained and the result is related to the surface area of the imprint.

Finally, with solid sludges, it is possible to measure the shrinkage, ie the reduction in volume as a result of drying ; in this way a distinction can be made between the solid state with shrinkage and the solid state without shrinkage.

VII. RELATIONSHIPS BETWEEN THE CHARACTERISTICS OF THE PHYSICAL STATE OF SLUDGES AND THE CHOICE OF MODE OF TRANSPORT, STORAGE, HANDLING AND SPREADING

Work has been carried out (2) to compare the results of tests for characterizing the physical state of sludges using simple equipment and the behaviour of these same sludges during handling and spreading tests.

The most suitable techniques are indicated in the following table.

Result of test to classify the sludge into one of the conventional physical states			Conventional physical state	Selection criteria	Most suitable techniques			
Flow test	Plasticity test	Shrinkage test			Transport	Stocking	Handling	Spreading
+	-	-	liquid	none, other than adequate concentration of solids (no stationary settling)	pumping or closed tanker	in pit	pumping	liquid-manure wagon
-	+	-	plastic	low mechanical stability (< 10)	trailer of closed skip	in pit	screw-drive pump or Archimedean screw	manure spreader or turbine spreader
-	+	-	plastic	mechanical stability (> 10)	lorry, skip or trailer	on concreted area	hydraulic fork with bucket	manure spreader or turbine spreader
-	-	+	solid	small particle size (< 5 cm)	lorry, skip or trailer	on concreted area	hydraulic fork with bucket or scraper	turbine spreader
-	-	+	solid	large particle size (> 5 cm) and depth of penetration > 5 mm	lorry, skip or trailer	on concreted area	hydraulic fork with bucket or scraper	spreader with moving bottom and sprocket wheel
-	-	+	solid	large particle size (> 3 cm) and depth of penetration < 5 mm	lorry, skip or trailer	on concreted area	hydraulic fork with bucket or scraper	previous maceration and turbine spreader

VIII. CONCLUSIONS

There are many methods available at the moment for characterizing the physical state of sludges.

These methods are clearly defined and are based on well-established scientific principles in the case of sludges capable of flowing under the force of gravity (liquid sludges) or under pressure (sticky or plactic sludges).

As regards solid sludges, we unfortunately have to be satisfied with relatively empirical and arbitrary tests of sludge behaviour. New methods for characterization undoubtedly remain to be discovered.

In all cases the means are available for a better choice and, where necessary, the design of storage, handling and transport equipment for sludges.

IX. ACKNOWLEDGEMENTS

The autor would like to express his gratitude in particular to the members of Working Party N° 1 who provided him with the information necessary for compiling this paper, and especially Dr Bruce.

X. BIBLIOGRAPHY

(1) F. Colin, Application of rheological techniques to the study of sewage sludges, La Tribune du Cebedeau No 317, pp. 178-187, April 1970.

(2) F. Colin and C. d'Ambrosio, Study of the characterization of the physical state of sludges (Study undertaken for the Ministry of the Environment, Waste Problems Dept.), Nancy, May 1978.

(3) W. Merkel, The flow characteristics of sewage sludge and other thick materials, Physics Vol 5, pp 355-361, November 1934.

(4) G. Annen, Calculation of the friction head losses of sewage-works sludges, Gas- und Wasserfach (Wasser), Vol. 104, No 12, pp. 342-346, March 1963

(5) V.C. Behn, Experimental determination of sludge flow parameters... (Translation into French: see (21)).

(6) V.C. Behn, Flow equations for sewage sludges, J.W.P.C.F. Vol 32, No 7, pp. 728-739, July 1960.

(7) R.I. Dick and B.B. Ewing, The rheology of activated sludge, J.W.P.C.F. Vol 39, No 4, pp. 543-560, April 1967.

(8) J.M. Sirman, Determination of sludge flow parameters, Water and Sewage Works, Vol 107, No 11, pp. 417-421, November 1960 (Translation into French).

(9) A.B. Metzner and J.C. Reed, Flow of non-Newtonian fluids, A.I.Ch.E. Journal, Vol 1, No 4, pp. 434-440, December 1955 (Translation into French).

(10) A. Doroszkowski and R. Lambourne, A viscosity-measuring technique for determining the thickness of polymer layers adsorbed onto titanium dioxide, Journal of Colloid and Interface Science, Vol 26, No 2, pp. 214-221, 1968 (Translation into French).

(11) A.S. Michaels and J.C. Bolger, Plastic flow behaviour of flocculated kaolin suspensions, Ind. Eng. Chem., Vol 1, No 3, pp. 153-161, August 1962 (Translation into French).

(12) R.J. Hunter and S.K. Nicol, Dependence relationships between the plastic flow behaviour of clay suspensions and surface properties, Journal of Colloid and Interface Science USA, Vol 28, No 2, pp. 250-259, October 1968 (Translation into French).

(13) W.D. Hatfield, The viscosity or pseudo-plastic properties of sewage sludge, Sewage Works Journal, Vol 10, No 1, pp. 3-25, January 1938.

(14) John R. Wolfs, Factors Affecting Sludge Force Mains, Sewage and Industrial Wastes, Vol 22, No 1, pp. 1-10, January 1950.

(15) Tsung-Lien and Chou, Resistance of sewage sludge to flow in pipes, Journal of the Sanitary Engineering Division, Proceedings of the ASCE Proc. Paper 1780, September 1958, SAS, pp. 1-19.

(16) V.C. Behn, Derivation of flow equations for sewage sludges, Journal of the Sanitary Engineering Division, Proceedings of the ASCE SA6, November 1960, pp. 59-81.

(17) C.F. Forster, The surface of activated sludge particles in relation to their settling characteristics, Water Research, Pergamon Press, 1968, Vol 2, pp. 767-776.

(18) Richard A. Bradley and Ray B. Krone, Shearing effects on settling of activated sludge, Journal of the Sanitary Engineering Division, Proceedings ASCE, SAI, February 1971, pp. 59-79.

(19) F. Colin et al, Characterization of sewage sludges (French work undertaken as part of European Action Project COST 68), Techniques et Sciences Municipales - L'Eau, No 1-76, January supplement 1976, pp. 2-23.

(20) A. Bousquet, Rheological study of a sludge, Institut de Mécanique des Fluides, Toulouse, 21 December 1977.

(21) V.C. Behn, Experimental determination of sludge flow parameters, Journal of the Sanitary Engineering Division, Proceedings of ASCE SA3, May 1962, pp. 39-54.

(22) H.E. Babitt and D.H. Caldwell, Laminar flow of sludges in pipes, Bulletin, University of Illinois Urbana Ill, Vol 37, No 12, 14 November 1939.

(23) H.E. Babitt and D.H. Caldwell, Turbulent flow of sludges in pipes, Bulletin Series, University of Illinois Urbana Ill, Vol 38, No 323, pp. 3-41, 13 November 1940.

(24) H.W. Campbell and P.J. Crescuolo, The use of rheoloy for sludge characterization, presented at the 11th IAWPR Conference, Cape Town 1982.

(25) M.C. Mulbarger et al, Pipeline friction losses for waste-water sludges, J.W.P.C.F., Vol 53, No 8, pp. 1303-1313.

(26) R.C. Frost, A method of estimating viscosity and designing pumping systems for thickened heterogeneous sludges, presented at "Hydro-transport 8", 8th International Conference on the Hydraulic Transport of Solids in Pipes, Johannesburg, South Africa, August 1982.

(27) A.P. Michaelson et al, Sewage sludge pumping - recent research and application, Water Pollution Control, 1982, pp. 220-240.

(28) Anonymous, The sludge viscometer you can put in the car, Water Research News, October 1981, p. 2.

(29) R.C. Frost, Prediction of friction losses for the flow of sewage sludges in straight pipes, Water Research Centre Technical Report 175, January 1982.

(30) F.E. Chavarria, Rheological study of biological floc suspensions, Doctorate thesis presented to Montpellier University, 9 June 1980.

(31) R.C. Frost, How to design sewage sludge pumping systems, Water Research Centre, W.R.C. Process Engineering, 85-S, September 1981.

(32) M. Reuss, Rheological properties of fermentation fluids, The Chemical Engineer, June 1982, pp. 233-236.

(33) J. Leadbitter and R.C. Frost, The design and use of a "Bucket Viscometer", Water Research Centre, W.R.C. Process Engineering 119-S, February 1982.

(34) D.J. Willis, A literature survey on sewage sludge pumping, B.H.R.A. TN 1500, October 1978.

(35) R.F. Wood and R.I. Dick, Some effects of sludge characteristics on dissolved air flotation, Progress in Water Technology, Vol. 7, No. 2, pp. 173-182, 1975.

(36) D.C.H. Cheng, R.W. Higman, "Design method for pipeline to handle non-newtonien fluids" Process Engineering pp. 19-24 - July 1969.

(37) R.R. Rimkus, R.W. Heil, "The Rheology of Plastic Sewage Sludge". Metropolitan Sanitary District of Greater Chicago.

(38) S.N. Bhattacharya; "Flow characteristics of primary and digested sewage sludge" Rheol. Acta. Vol. 20, pp. 288-298 - 1981.

(39) G. Sarmiento et al., "Measurement of the rheological characteristics of slowly settling floculated suspensions". Ind. Eng. Chem. Process. Des. Dev. Vol. 18, No 4, pp. 746-751 - 1979.

(40) D.S. Parker et al., "Physical conditioning of activated sludge floc" J.W.P.C.F., Vol. 43, No 9, pp. 1817-1833 - sept. 1971.

(41) E. Luggen, "Charakterisierung von Klaerschlamm Aufgrund des Fliess-verhaltens und der Teilchengroesseverteilung". Thèse de Doctorat, EPF Zurich, DISS ETH 5740, 1976.

(42) A. Geinopolos, J. Katz, "A study of a Rotating Cylinder Sludge Collector in the Dissolved-Air Flottation Process". Journal WCPF, June 1914, Vol. 36, No 6, pp. 712-721.

(43) F. Colin, "Détermination de l'humidité minimale et des caractéristiques physiques limites des substrats boueux admissibles en méthanisation". Communication au Séminaire Méthanisation du COMES Sophia-Antipolis - 10-11 juin 1982.

(44) D. Bellet, C. Thiriot, "Mesures de divers paramètres rhéologiques" La Houille Blanche, Mai 1970, pp. 415-426.

(45) M. Martin, "Caractéristiques rhéologiques et composition des boues de forage" La Houille Blanche, mai 1970, pp. 463-470.

(46) R.C. Frost, Communication personnelle - WRC Stevenage - 15.7.1982.

(47) A. Geinopolos, W.J. Katz, "A study of a rotating cylinder sludge collector in the dissolved-air flottation process", J.W.P.C.F., Vol. 36, No 6, pp. 712-721, June 1964.

(48) D.J. Willis and G.F. Truscott, A survey of solids-handling pumps and systems, BHRA Report No TN 1363, 55 pages (May 1976).

(49) A.P. Fisichelli, Raw sludge pumping - problems and interdisciplinary solutions, Water Pollution Control Journal, 42, 11, pp. 1916-1921 (1970).

(50) G.E. Hubbell, Dearborn sewage sludge pumped through long pipeline, Engineering News Record, 111, p. 343, 21 September 1933.

(51) D.J. Lambert, Pump applications in sewage sludge dewatering process, Pumps for Progress - 4th Technical Conference of BPMA (organized with BHRA) Durham, UK Paper C2, pp. 77-88 (April 1975).

(52) U.T. Mann, Centrifugal slurry pumps for handling municipal sewage sludge, Water and Sewage Works, 114, 2, pp. 52-53, February 1967.

(53) R.D. Palmer, Beyond the bend - Allen Gwynnes sewage pumps, A.P.E. Engineering, 19, pp. 9-17, July 1975.

(54) Anonymous, Procedures adopted by the Central Laboratory for Bridges and Highways, Atterberg, Mode opératoire SI 4-1963, pp. 4-15, Dunod, Paris 1965.

DISCUSSION

H.W. Campbell

It seems to me open to question as to whether liquid sludges have a definite yield point. Our rheogram curves for liquid sludges all go through the origin, indicating a zero yield strength. In the work of Wood and Dick, yield stresses are given but they extrapolated rather than measured values. Liquid sludge does not seem to translate into a measureable yield value.

F. Colin

There are many difficulties in characterising the rheology of well-flocculated sludges, especially with rotating viscometers in which such sludges separate into a shell of water and a solids layer, resulting in unrepresentative rheological measurements.

G. Hoyland

Do you have any information on the influence of temperature on the values of rheological constants for sludges?

No. The influence of temperature is well known for water but not for sludges.

T.J. Casey

Dr. Colin has touched on the problem of making rheological measurements on well-flocculated sludges using standard rotating viscometers. It would seem to me that phase separation must be inevitable to some degree with sludges, depending on the extent of conditioning, in such viscometers. How reproducible, therefore, are measurements of properties such as viscosity?

Reproducibility is good for poorly flocculated sludges but not good for well-flocculated sludges. We have developed a new measuring device to overcome the problem of phase separation. It consists of a rotating paddle in a vessel having a fixed counter paddle system. This arrangement effects dispersion of the mixing energy throughout the sludge mass. By measuring the applied torque, the power input can be evaluated.

R.C. Frost

In addition to the use of mixing viscometers of the type referred to by Dr. Colin, there is another procedure by which the problem of phase separation can be overcome, while still using concentric cylindrical viscometers. By carrying out measurements at various fixed speeds for finite time periods

Yes, time is a very important parameter and we neglect it in making such measurements. It is difficult to take it into account with rotating viscometers, but it can be taken into account with tube viscometers by using tubes of fixed diameter but differing length.

and extrapolating back to zero time at each speed, it is possible to generate the laminar flow curve at zero time.

INTER-RELATION BETWEEN SLUDGE CHARACTERISTICS

R.C. FROST
Water Research Centre

Summary

Evidence is provided in this paper for the existence of intimate associations between certain aspects of rheological and solids-liquid separation behaviour. The potential value of these associations for sewage sludge treatment design and management lie in three areas.

Firstly in the use of easily measured thickening characteristics to provide a better estimate of rheological properties than is provided by solids concentration data. Secondly, the associations may be used as a basis for a sludge consistency scale, which could assist the evolution of integrated design approaches. Lastly, knowledge of their existence may stimulate more fundamental work aimed at determining the mechanisms whereby particle-fluid interactions control sludge behaviour.

1. INTRODUCTION

Other papers presented at this workshop on sewage sludge characteristics deal with particular methods of characterisation, with the view of aiding the design and management of sludge treatment and disposal systems. In contrast, this paper addresses itself to the inter-relations that exist between some characteristics, and attention is focused upon the dewatering and rheological properties of sludges. It is appropriate to state initially the benefits that might be gained from such a comparative study, and it may be helpful to consider these benefits as being of a short, medium, and long term nature.

In the short term the benefit to be gained from a knowledge of the inter-relation between two characteristics is essentially of a practical nature. If the measurement of one of these properties presents relatively greater difficulty, then determination of the more easily measured property can be used to estimate that which is desired. This approach is especially useful if the required property is solids concentration dependent, but can be estimated at any concentration from measurements of the associated property made at another concentration. A good example of this will be shown to be the estimation of the rheological properties of thickened sludges from their more easily measured thickening properties (determined using sludge at normally available solids concentrations).

In the medium term a knowledge of the inter-relation between sludge characteristics may assist the evolution of integrated approaches to design, so assisting the optimum selection of process equipment, combination of operations, and system design and operation. A speculative illustration of an integrated design approach using a sludge consistency scale will be presented in the paper. This scale is based on the relationship between rheological and dewatering properties; the degree of solids-liquid separation achieved by different equipment; and the speculated limitations imposed by the rheological characteristics of a sludge on both the suitability for use of subsequent process equipment, and the ease of assimilation of sludge disposed of in the environment.

A long term goal in any field of technical endeavour is the ability to predict from theoretical considerations the observed macroscopic behaviour of a system - whether the system be a material whose characteristics we wish to know or a process or operation to which material is applied. Empirical relations between sludge characteristics may help in this effort if their existence stimulates the work needed to understand the fundamental mechanisms by which particle and liquid properties interact, and are manifested as observed sludge behaviour. A discussion of some possible explanations for the observed connexion between rheological and dewatering properties of sludges is offered.

2. METHODS OF SLUDGE CHARACTERISATION CONSIDERED

2.1. DEWATERING

2.1.1. Settleability

The rate and degree of settling of activated sludge suspensions at solids concentrations representative of those prevailing in typical mixed liquors have been assessed using the standard SSVI technique[1]. Values of SSVI are reported for initial suspended solids concentrations of 3500 mg/l.

2.1.1. Thickenability

The thickenability of a wide range of sludges was measured using a slow speed, frozen image centrifuge developed at the Water Research Centre[2,3]. By itself this test does not comprehensively characterise the thickening behaviour of a sludge, as has been shown by Hoyland and Day[4], but the relationship gained between the nominal compressive pressure, NCP, and the ultimate average suspended solids concentration, $\bar{C}_\infty$, is adequate for the present purposes. This relationship is conveniently defined by equation 1.

$$\bar{C}_\infty = a \ NCP^b \qquad (1)$$

where $NCP = C_o \ h_o \ g \qquad (N/m^2)$

C_o = initial suspended solids concentration (kg/m^3)

h_o = initial sludge blanket depth (m)

g = centrifugal acceleration (m/s^2)

a, b = empirical coefficients characterising a sludge

2.1.3. Filtrability

Filtrability was assessed for a number of raw and anaerobically digested sludges using capillary suction time measurements[5]. To be useful in comparing different sludges the measurements should be made at the same solids concentration, and the data are expressed in the main for a suspended solids concentration of 33 kg/m^3.

A piston press technique[6] has been used also, to extend to higher compressive pressures the information yielded by centrifuge testing. The results of the piston pressing tests can be expressed by equation 2, analagous to equation 1.

$$C_\infty = a^1 \, CP^{b^1} \qquad (2)$$

where C_∞ = ultimate solids concentration of a pressed cake after 'infinite' time (kg/m^3)

CP = externally applied compressive pressure (N/m^2)

a^1, b^1 = empirical coefficients characterising a sludge

2.2. RHEOLOGY

2.2.1. Steady state flow behaviour

The laminar flow behaviour of sludges was determined under steady state conditions using a wide bore tube-viscometer, having tubes of 12 mm, 19 mm and 27 mm diameter[7,8,9]. Use of this instrument yields paired readings of pressure drop and flow rate, which can be transformed to provide a relationship between shear stress and shear rate[8]. Herschel-Bulkley fluid model parameters can normally be found to describe the laminar flow curve, this model being defined by equation 3.

$$\tau = \tau_y + K \gamma^n \qquad (3)$$

where τ = shear stress (N/m^2)

τ_y = yield stress or static resistance (N/m^2)

γ = shear rate or rate of strain (s^{-1})

K = consistency coefficient (N.s^n/m^2)

n = consistency index

Methods of using the information gained for the design of pumping systems are considered in detail elsewhere[9.10].

2.2.2. Cake plasticity

The plasticity or malleability of sludge cake has been determined using a Seta (Universal Model) penetrometer. To the cone attachment (IP 50 and ASTM 217) was added a mass of 0.15 kg. The penetration depths recorded are a purely empirical measure of cake plasticity.

2.3. RESULTS

The results of the characterisation tests stated above are listed with those pertaining to other, more general, tests in Tables 1, 2 and 3.

Table 1. Dewatering and general characteristics of a range of sludges

Sludge source and type	Loss on ignition of suspended solids (wt%)	Grease content of volatile solids (wt%)	Particle size distribution (wt%)* volatile fines	non-volatile fines	volatile coarse	$SSVI_{3500}$ (ml/g)	Thickenability and dewaterability coefficients a ($kg^{1-b}.s^{2b}/m^{3-b}$)	b	a' ($kg^{1-b'}.s^{2b'}/m^{3-b'}$)	b'	CST_{18} (s) after 40s shear at a solids concn of ( ) kg/m^3
A primary	82.4	36.2	28.0	9.9	54.3	-	12.4	0.275	-	-	175 (33)
B "	47.7	26.3	30.2	47.7	17.5	-	40.3	0.224	-	-	77 (33)
C1 "	54.7	23.0	32.3	35.6	22.4	-	42.1	0.155	-	-	93 (33)
D "	76.9	-	42.9	17.1	34.0	-	22.0	0.204	-	-	-
E1 " + humus	68.6	25.3	(			-	30.7	0.192	-	-	
E2 " "	75.9	31.4	(23.3	17.9	49.1	-	20.0	0.228	-	-	205 (33)
E3 " "	72.7	28.3	(			-	21.3	0.221	-	-	
F " "	75.2	18.6	13.7	12.6	61.5	-	19.1	0.213	-	-	195 (33)
G " "	73.2	22.5	31.5	12.8	41.6	-	31.2	0.151	-	-	315 (33)
H " "	70.9	-	-	-	-	-	20.1	0.225	-	-	-
I1 " "	74.5	29.7	46.6	21.4	27.9	-	21.3	0.224	-	-	350 (33)
J " "	65.9	-	-	-	-	-	22.8	0.200	-	-	-
K " "	85.5	22.2	50.7	11.3	34.8	-	13.3	0.241	-	-	480 (33)
L " "	81.5	16.7	40.4	12.1	41.1	-	19.2	0.170	-	-	350 (33)
M " "	77.7	18.7	48.0	17.5	29.7	-	18.0	0.200	-	-	400 (33)
N " "	69.7	20.7	-	-	-	-	24.3	0.178	-	-	-
O " "	81.6	-	-	-	-	-	8.5	0.301	-	-	-
P " "	-	-	-	-	-	-	-	-	28.7	0.222	-
C2 humus	65.8	-	-	-	-	-	29.2	0.128	-	-	-
Q1 nitrifying, partly	-	-	-	-	-	31	-	-	-	-	-
Q2 denitrifying activated	-	-	-	-	-	40	-	-	-	-	-
Q3 " "	85.2	-	-	-	-	-	10.4	0.233	-	-	-
R1 nitrifying activated	79.2	2.2	-	-	-	-	14.1	0.178	-	-	-
R2 " "	-	-	-	-	-	59	-	-	-	-	-
R3 " "	-	-	-	-	-	68	-	-	-	-	-
S non-nitrifying "	86.0	-	-	-	-	122	9.46	0.308	-	-	-
T activated	-	-	-	-	-	53	-	-	-	-	-
I2 anaerobic digested	57.7	14.8	45.3	32.2	17.3	-	21.6	0.209	-	-	165 (33)
U1 anaerobic digested	61.9	11.3	(			-	28.8	0.137	-	-	600 (37.6)
U2 " "	61.3	8.9	(42.4	32.2	19.6	-	15.2	0.226	-	-	
V " "	66.6	48.1	-	-	-	-	41.9	0.154	-	-	-
W " "	64.6	-	-	-	-	-	20.7	0.172	-	-	-
X " "	54.9	-	-	-	-	-	37.8	0.188	97.4	0.158	-
Y coal waste slurry	-	-	-	-	-	-	-	-	284	0.128	-

* fines defined as material passing a 250 µm sieve

Table 2. Herschel-Bulkley fluid model parameters

Sludge	Suspended solids concn (kg/m^3)	Temperature (°C)	τ_y (N/m^2)	K (N.s^n/m^2)	n
A	76.8	9.8	0.0	4.20	0.363
A	60.8	9.8	1.50	0.948	0.511
A	41.0	10.5	0.716	0.114	0.725
B	113.4	10.0	0.0	1.42	0.562
B	90.1	10.0	0.0783	0.592	0.528
B	55.9	9.5	0.0	0.150	0.573
C1	126.6	3.2	2.32	1.93	0.563
C1	53.1	3.1	0.176	0.0837	0.671
C1	84.6	3.1	0.467	0.433	0.699
C1	135.2	8.8	0.0	3.67	0.554
C1	112.0	9.5	3.17	0.734	0.666
D	57.9	9.5	0.0	1.10	0.475
D	57.8	11.0	0.0	0.995	0.450
D	53.7	10.5	0.259	0.668	0.480
D	43.0	11.5	0.455	0.164	0.635
D	60.6	12.2	0.618	0.330	0.556
D	32.6	12.0	0.0	0.185	0.443
D	22.5	10.2	0.252	0.00987	0.864
E1	99.0	19.0	0.0	2.04	0.480
E1	75.2	20.0	0.543	0.226	0.768
E1	47.4	20.6	0.0	0.169	0.599
E2	48.9	16.5	0.315	0.189	0.673
E2	31.0	16.0	0.207	0.051	0.735
E3	39.4	17.1	0.386	0.151	0.664
F	66.2	9.6	4.41	0.317	0.895
F	53.3	9.7	0.414	0.728	0.628
G	73.6	7.5	2.75	0.920	0.693
G	69.8	7.4	0.0	1.60	0.561
G	33.4	7.3	0.205	0.166	0.628
G	52.8	4.8	0.548	0.507	0.624
H	57.0	4.8	0.0	0.998	0.517
I1	23.7	21.4	0.0	0.0083	0.514
I1	41.4	17.8	0.314	0.195	0.594
I1	66.4	24.0	0.910	1.01	0.503
J	33.4	16.0	0.695	0.182	0.568
J	20.8	16.0	0.0	0.120	0.475
K	47.7	4.6	0.0	3.54	0.397
K	32.8	4.8	0.284	0.690	0.513
K	20.5	5.6	0.254	0.111	0.624
L	45.9	2.0	0.0	3.32	0.460
L	33.5	2.0	1.76	0.389	0.636
L	26.8	2.2	0.733	0.201	0.649
M	29.1	1.1	0.733	0.283	0.585
M	20.3	2.4	0.284	0.104	0.631
M	37.0	0.8	0.599	1.06	0.471
N	62.0	8.6	5.06	1.78	0.484
O	92.7	11.5	0.0	18.0	0.452
O	84.9	12.9	32.6	6.00	0.527
C2	47.0	6.5	2.77	0.509	0.607
C2	32.8	7.5	0.321	0.213	0.526

Table 2 continued

Sludge	Suspended solids concn (kg/m^3)	Temperature (^oC)	τ_y (N/m^2)	K ($N.s^n/m^2$)	n
Q1	27.9	9.0	0.399	0.0173	0.866
Q2	22.7	11.0	0.422	0.0294	0.797
Q3	23.6	3.5	0.485	0.211	0.600
Q3	20.6	3.0	0.405	0.163	0.590
Q3	33.4	12.2	1.02	1.44	0.460
Q3	22.3	12.2	0.875	0.0716	0.708
Q3	17.6	10.8	0.274	0.0709	0.626
R1	43.3	15.0	21.3	1.60	0.571
R1	35.8	13.5	8.11	0.729	0.594
R1	32.4	14.5	7.82	0.0588	0.921
R1	28.3	13.5	3.33	0.127	0.734
R1	24.0	13.0	1.27	0.141	0.672
R1	25.2	12.0	1.42	0.270	0.589
R1	20.2	13.0	0.785	0.0527	0.752
R1	11.8	12.5	0.155	0.0440	0.653
R2	19.3	8.2	0.345	0.0604	0.679
R2	24.8	8.5	0.760	0.211	0.565
R2	29.4	9.6	1.53	0.503	0.492
R2	31.7	10.0	2.25	0.542	0.513
R3	18.0	10.3	0.879	0.0398	0.747
R3	21.4	8.8	1.13	0.176	0.578
S	19.2	3.5	0.682	0.380	0.495
S	12.8	2.0	3.41	0.0939	0.598
T	27.3	15.0	0.539	0.124	0.674
T	18.8	15.0	0.246	0.0187	0.800
I2	25.9	18.3	0.0	0.211	0.327
I2	27.6	16.0	0.215	0.0979	0.502
I2	43.3	17.0	0.491	0.220	0.581
I2	55.8	17.3	1.92	0.491	0.493
I2	64.9	18.7	2.53	0.992	0.439
I2	71.7	18.5	4.37	1.27	0.458
U1	76.3	5.0	0.0	15.5	0.224
U1	58.6	4.0	4.80	1.32	0.533
U1	41.1	4.8	1.41	0.296	0.602
U2	88.2	10.3	36.7	8.78	0.457
U2	83.4	8.1	33.6	5.59	0.507
V	138.0	7.3	14.8	1.91	0.612
V	111.8	8.3	4.45	0.610	0.638
V	58.7	9.3	0.429	0.0541	0.733
W	31.3	8.7	0.561	0.217	0.472
W	46.1	8.3	2.26	0.243	0.669
W	61.5	7.7	0.783	0.887	0.581

Table 3. Results of penetrometer tests

Sludge	Suspended solids concentration (kg/m^3)	Penetration (mm)
P	150	30.2
P	182	23.4
P	228	16.0
P	238	22.0
P	240	21.0
P	325	8.8
P	334	10.3
P	357	6.3
P	384	6.2
P	396	6.4
P	397	6.0
P	450	4.4
P	482	3.8
P	550	3.0
P	610	2.3
X	622	3.4
X	806	1.5
Y	1015	22.0
Y	1120	18.8
Y	1207	10.6
Y	1237	9.6
Y	1502	4.1
Y	1614	2.1

3. THE EVIDENCE FOR INTER-RELATIONS BETWEEN CHARACTERISTICS

3.1. SETTLEABILITY AND RHEOLOGY

Over a number of years, Dick and coworkers[11,12] investigated the rheology of activated sludge within the range of solids concentration confirming to that of mixed and return liquors. Experiments were conducted in which activated sludges differing in physical properties were produced by varying the food to microorganism and carbon to nitrogen ratios. This resulted in sludges whose composition ranged from the zoogleal types commonly found in activated sludge processes to those comprised almost entirely of filamentous growths - of which three distinct types were produced. Rheological and batch flotation tests were made on these sludges, and it was reported[11] that the rate of rise and the float concentration were both inversely related to sludge viscosity at a reference solids concentration.

In a study to obtain rheological data for the design of a 5.5 km pipeline, Rose-Innes and Nossel[13] observed that the relationship between viscosity and solids concentration was not unique for different activated sludges. At any one solids concentration the viscosity of activated sludge showed a broad trend to increase with increasing settled volume index SVI.

To confirm these various observations, data from Tables 1 and 2 for sludges Q1, Q2, R2, R3, S and T have been used to prepare a correlation between an apparent viscosity of sludge at a solids concentration of 30 kg/m^3 and the corresponding $SSVI_{3500}$ value, see Fig. 1. It is clear from equation 4 that the shear rate at which apparent viscosity is reported has to be defined. For the purpose of Fig. 1 a shear rate of 100

$$\text{apparent viscosity} \qquad \eta = \frac{\tau_y + K.\gamma^n}{\gamma} \qquad (4)$$

reciprocal seconds has been chosen, this being characteristic of a number of situations. The apparent viscosity at $\gamma = 100\ s^{-1}$ can thus be termed the characteristic viscosity η_c of a sludge.

Inspection of Table 2 shows that a value for η_c at 30 kg/m^3 could not always be interpolated from the data. In those cases η_c values were extrapolated with respect to solids concentration on the basis that viscosities tend to increase exponentially with solids concentration, with an exponent value of around 3.

3.2. THICKENABILITY AND RHEOLOGY

As stated above Wood and Dick[11] made observations of an inverse relationship between float solids concentration and sludge viscosity at a reference solids concentration. Other links between thickener performance and sludge rheology have been observed by Geinopolous and Katz[14], who related the capacity of the collector for a dissolved-air flotation unit to the rheology of the collected sludge. However, though interesting as information, these observations lack formality and relate only to activated sludge.

To examine whether there might be a more general link between thickening and rheological properties, a survey was made of different sewage sludges. These sludges were chosen so as to display a wide range of

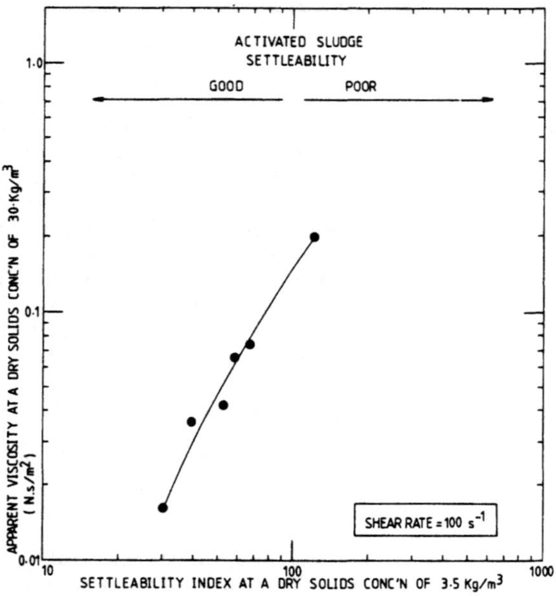

Fig. 1. Inter-relation between activated sludge settleability and rheology

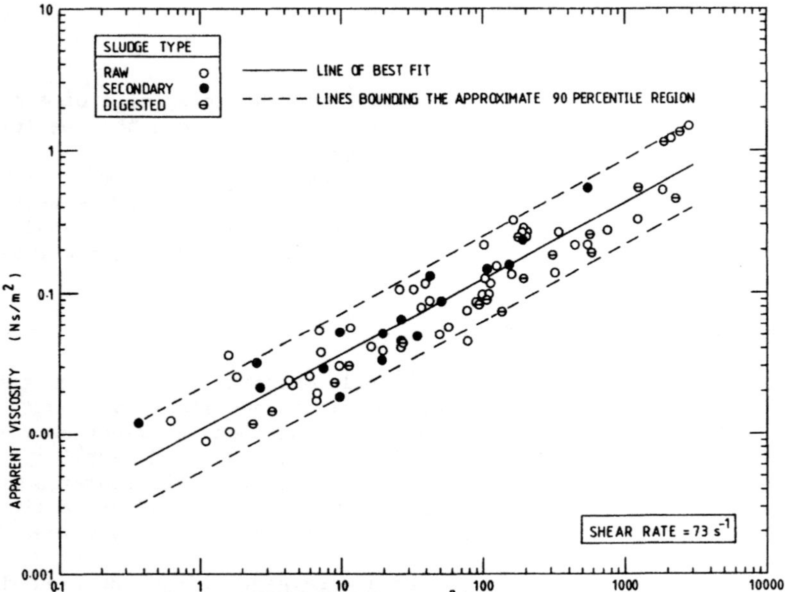

Fig. 2. Inter-relation between thickenability and rheology

thickenability and viscosity. To appreciate the degree of diversity that can exist it will be useful to note that the viscosity of these sludges would, at any one solids concentration, cover a 100 fold range.

It was hypothesised that the rheological state of any sludge could be defined approximately by the minimum nominal compressive pressure required to produce the sludge. In other words, the application of a given nominal compressive pressure to different sludges would, at the completion of thickening, result in thickened sludges of similar viscosity - though their solids content could be quite different. If the rheological characteristics of a sludge refer to a solids content of $\bar{C}_\infty$, then the appropriate value of NCP according to this hypothesis is found by recasting equation (1):

$$NCP = \left(\frac{\bar{C}_\infty}{a}\right)^{1/b} \qquad (5)$$

This hypothesis was tested using the data listed in Tables 1 and 2, apparent viscosities being compared with the NCP values calculated using equation (5). Comparisons have been made over the shear rate range 1 to 2180 reciprocal seconds[10], the relation shown in Fig. 2 being for a shear rate of 73 s^{-1} at which peak correlation of the data was observed. No significant reduction of the residual variance could be made by taking into account variations in the loss on ignition of the different sludges or the temperatures at which rheological measurements were made.

Correlating the viscometric data in this way, with nominal compressive pressure replacing solids concentration as the correlating parameter, results in the variation of viscosity with the correlating parameter being reduced from a factor of 100 to one of 4.

3.3. DEWATERABILITY AND RHEOLOGY

An analagous hypothesis to that above has been tested at greater degrees of solids-liquid separation using piston pressing and penetrometer data abstracted from Tables 1 and 3. Information for sludges P, X and Y has been used to plot Fig. 3, which shows there to be a relationship between the rheological state of a sludge cake and the minimum pressure which must be applied to form the cake.

Capillary suction time and characteristic viscosity data have also been compared, at a sludge solids concentration of 33 kg/m^3. However, only a very broad trend for η_c to increase with increasing CST was observed, and the results are not reproduced here.

3.4. ASSOCIATION OR CAUSE AND EFFECT

Wood and Dick[11] appeared to believe that the rheological properties of a sludge played a direct role in determining the solids concentration of sludges produced by thickening plant. However, though the performance - as assessed by the solids content of the output - of solids - liquid separation plant at full scale may, in principle, be influenced by the rheological properties of the sludge involved, it may be a mistake to regard the above empirical relations (3.1-3.3) as being other than mere associations between parameters. There is no strong case for considering them to reflect a cause and effect relation.

On this view, the cause and effect relationships occur at a much deeper level, and would describe the mechanisms whereby particle characteristics - alone and in interraction with fluid characteristics - determine the gross behaviour of the aggregate sludge. Figure 4 has been drawn to illustrate this conception. The positive aspect of this view is that there is a high probability that the fundamental mechanisms determining the rheological and solids-liquid separation behaviour of sludges have much in common.

4. USE OF INTER-RELATIONS FOR HYDRAULIC DESIGN

The measurement of a sludge's viscometric properties is not a simple matter. Rather, direct viscometric measurements and their processing and interpretation require staff possessing relatively specialised skills of quite a high order - skills which for several historical reasons are not commonly found in the sewage treatment industry. The situation is exacerbated by the unfortunate fact that, without modification, much of the commercially available equipment is not suitable for making absolute measurements. There is a role to be played, therefore, by a simple indirect method of estimating sludge viscosity.

Once calibrated a 'bucket-viscometer'[15,16] can be used by staff at almost any level of competence: the measurement is simple and takes minutes only, and taking a reading from a calibration chart is the only data 'processing' that is required. However, its use will tend to be limited to the determination of sludge viscosity at existing solids concentrations, and a major requirement in practice is the estimation of viscosity of thickened sludges - before they have been thickened.

To meet this requirement laminar flow curves have been simulated as a function of nominal compressive pressure, using information of the type displayed in Fig. 2[9,10]. Hence the relatively easy measurement of sludge thickenability, made using a small sludge sample, can provide an estimate of the rheological properties of the sludge at almost any solids content. Table 4 lists some conservative values of the rheological parameters τ_y, K and n as functions of NCP.

Table 4. Conservative values of Herschel-Bulkley model parameters as a function of nominal compressive pressure

NCP (N/m^2)	τ_y (N/m^2)	K ($N.s^n/m^2$)	n
0.1	0.102	0.0168	0.710
1.0	0.385	0.0624	0.682
10	1.47	0.223	0.660
100	5.53	0.834	0.630
1000	20.9	3.05	0.604
10000	79.7	10.7	0.581

The hydraulic design procedure which would make use of this information is outside the scope of this paper. Interested readers should consult References 8, 9 and 10.

5. A SLUDGE CONSISTENCY SCALE

A potentially powerful, extensive use of the link between rheological and dewatering properties, is as an aid to identifying optimum sludge concentrations for specific treatment and disposal routes, and vice versa. A sludge consistency scale can be developed - analagous to the pH scale -

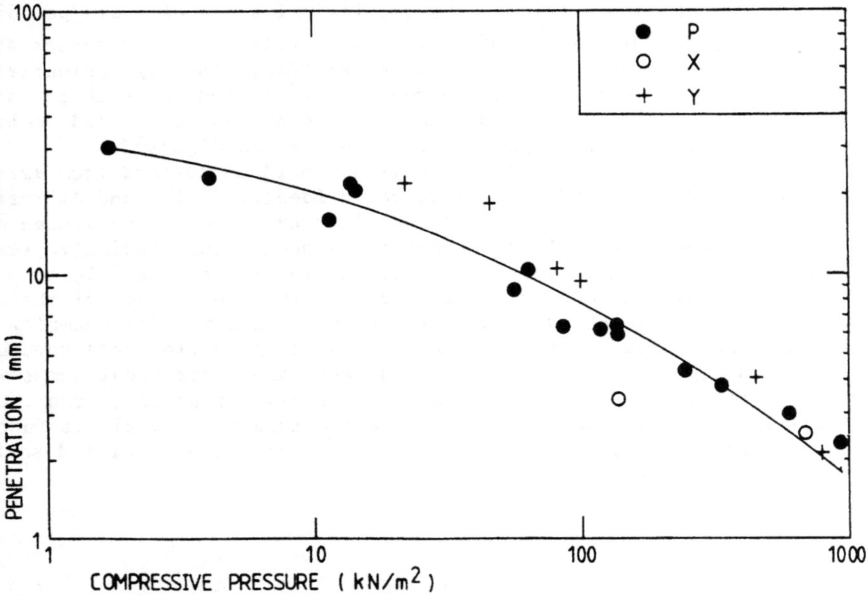

Fig. 3. Inter-relation between dewaterability and rheology

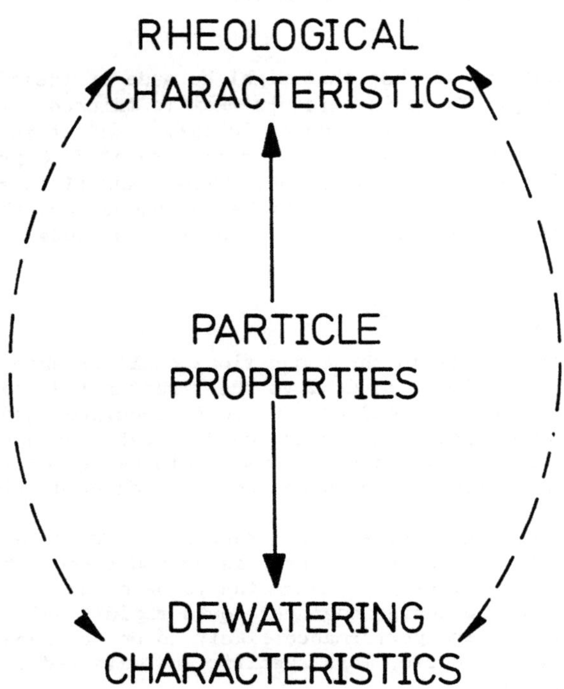

Fig. 4. A conceptual link between dewatering and rheological characteristics

the value of which would indicate the rheological state of a sludge. Since the reliability and efficiency of many unit operations and processes depen on the rheological properties of the material being handled, appropriate band widths on the consistency scale could be allocated for each operation Other material properties are important too, so scales would need to be drawn for each main sludge type - raw, secondary and digested.

Scale positions for specific operations could be defined from agreed process design criteria and from empirical evidence. Colin and Ambrosio[1 for example, have related the ease and efficiency of spreading sludge on land to easily measurable sludge properties - mechanical stability, resistance to penetration, and adhesivity. It should also be possible to characterise dewatering operations in terms of the consistency of the sludges produced. Process design criteria could involve, for example, the requirement that turbulent flow must exist (at a given flow rate through a given sized pipe) in order that heat exchangers and centrifugal pumps woulc operate efficiently. Hence noting that compressive pressure (formation pressure) and nominal compressive pressure are related by a simple functior of the dry solids and fluid densities, see equation 6, a completed scale might take the form of Table 5.

$$CP = \left(\frac{\rho_s - \rho_w}{\rho_s}\right) NCP \qquad (6)$$

where ρ_s = effective density of dry matter
ρ_w = density of water

As an example of the use that might be made of these scales, consider the case following. There is a requirement to thicken a raw sludge up fron 30 kg/m^3 to a solids concentration of 80 kg/m^3, and it is then proposed that the sludge be pasteurised at a temperature of 70°C prior to anaerobic digestion. A feasibility study is being made, and it is especially uncertain as to how the sludge will be thickened, pumped and then pasteurised. If thickening tests were made on a sample of this sludge, and these yielded

$$\bar{C}_{\infty} = 19.2\ NCP^{0.17}$$

then provisional answers to these questions could be supplied by reference to scales such as Table 5: to achieve the target solids concentration of 80 kg/m^3 a minimum NCP of 4420 N/m^2 would be required, and centrifugation or polymer assisted gravity drainage would probably be needed to produce the sludge; a positive displacement pump would be needed to pump the thickened sludge; and pasteurisation would be best achieved using steam injectio as opposed to a heat exchanger.

By characterising materials and processes in this way, therefore, the preliminary stages of design of a processing route could be much assisted. The scientist could better appreciate the value of any test results he may have obtained, and the design engineer could rapidly gain a feel for the equipment needed and the performance likely to be obtained. At further stages of design, of course, more specific and detailed information would usually be required.

Table 5. A speculative sludge consistency scale

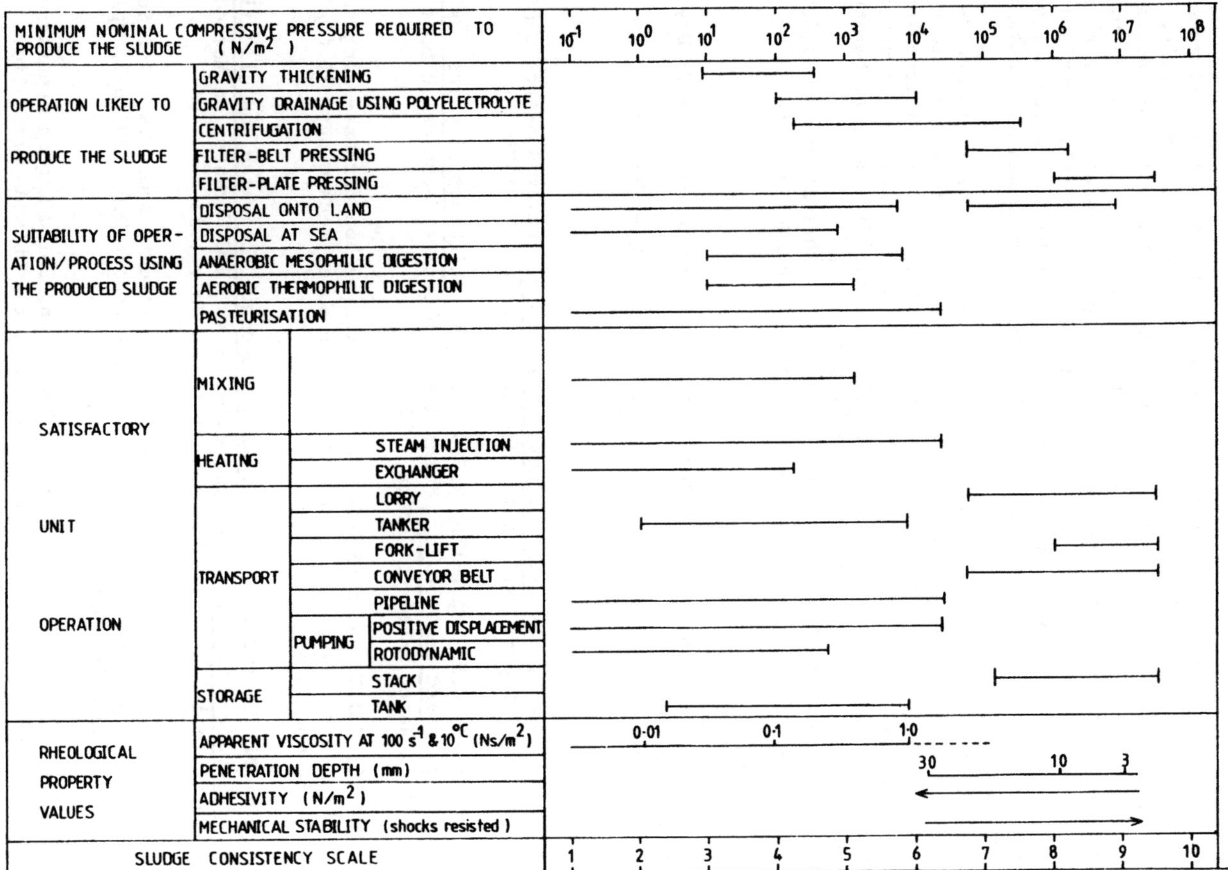

6. FUNDAMENTAL CONSIDERATIONS

An explanation for these interrelations may lie in the influence of the bound water contents of sludges. It is well known, for example, that the viscosity of a suspension of rigid, neutrally buoyant, non-interactive spheres is solely dependent on the volumetric particle concentration. It might seem reasonable to suppose, also, that particles possessing a greater bound water content would compress down to a lower dry solids content. An interpretation of these results, therefore, might be that under a given compressive pressure sludges will thicken to an approximately constant volumetric concentration of particles.

A weakness with this idea is that it begs questions as to the definition of a particle, and their properties. It also ignores the nature of the forces which have to be overcome to achieve compressive thickening, and ignores the many other mechanisms whereby sludge rheology is determined. Particle morphology and structure, elasticity, rigidity may all be important factors, as might their electrical properties.

Much fundamental work remains to be done. If this paper can assist in stimulating the effort required then it will have proved useful.

ACKNOWLEDGEMENTS

Piston press and penetrometer tests were carried out by M.C. Day and R.C. Baskerville.

This paper is published by permission of the Director, WRC.

REFERENCES

1. WHITE, M.J.D. Settling of activated sludge. Water Research Centre Technical Report TR11, Water Research Centre, UK, 1975.

2. LOCKYEAR, C.F. and WHITE, M.J.D. The WRC thickenability test using a slow-speed centrifuge. Water Research Centre Technical Report TR118, Water Research Centre, 1979.

3. WHITE, M.J.D. and LOCKYEAR, C.F. A novel method for assessing the thickenability of sludges. Progress in Water Technology, 10 (5/6), 1978, pp. 33-43.

4. HOYLAND, G. and DAY, M.C. Design procedure for batch consolidation of sewage sludges. Paper presented to the EEC Workshop on Methods of Characterising Sewage Sludges, Dublin 6 July 1983.

5. GALE, R.S. Optimising the use of pretreatment chemicals. In: Solid/Liquid Separation Equipment Scale-up, Purchas, D.B. (Editor), Uplands Press Ltd., Croydon, UK., 1977, pp. 39-80.

6. PURDY, J.A. Variable volume filters. In: Solid/Liquid Separation Equipment Scale-up, Purchas, D.B. (Editor), Uplands Press Ltd., Croydon, UK, 1977, pp. 430-443.

7. JOHNSON, M. First report on the WRC sewage sludge pumping project. Water Research Centre Technical Report TR162, April 1981, 102 pp. Water Research Centre and BHRA Fluid Engineering, UK.

8. FROST, R.C. *Prediction of friction losses for the flow of sewage sludges along straight pipes*. Water Research Centre Technical Report TR175, Water Research Centre, UK, 1982.

9. FROST, R.C. *How to design sewage sludge pumping systems*. Water Research Centre Technical Report TR185 Water Research Centre, 1983.

10. FROST, R.C. and OVENS, J.A. A method of estimating viscosity and designing pumping systems for thickened heterogeneous sludges. *Proceedings of Hydrotransport 8*, 8th International Conference on the Hydraulic Transport of Solids in Pipes, Johannesburg, South Africa. August 25-27, 1982. Published by BHRA Fluid Engineering, Cranfield. Paper P1, p 485-501.

11. WOOD, R.F. and DICK, R.I. Some effects of sludge characteristics on dissolved air flotation. *Progress in Water Technology*, *7* (2), 1975, pp. 173-182.

12. DICK, R.I. and EWING, B.B. The rheology of activated sludge. *Journal Water Pollution Control Federation*, *39*, (1967), 543-560.

13. ROSE-INNES, I.H. and NOSSEL, S. *The rheology and pumping of thickened activated sludge.* Paper presented to the eleventh conference of the IAWPR post conference seminar on modern trends in sludge management, 5-6 April 1982, Pretoria, South Africa.

14. GEINOPOLOS, A. and KATZ, W.J. A study of a rotating cylinder sludge collector in the dissolved air flotation process. *Journal Water Pollution Control Federation*, *36* (6), (1964), 712.

15. LEADBITTER, J. and FROST, R.C. *The design and use of a bucket-viscometer*. Water Research Centre Laboratory Record 119-S, February 1982, Water Research Centre, UK.

16. COLIN, F. *Characterisation of the physical state of sludges*. Paper presented to the EEC Workshop on Methods of Characterising Sewage Sludges, Dublin 6 July 1983.

17. COLIN, F. and AMBROSIO, G.d'. *Study of the characterisation of the physical state of sludges*. (Etude de la characterisation de l'etat physique des boues). Institut de Recherches Hydrologiques, Nancy, France, May 1978, 97pp. (In French).

DISCUSSION

G. Mininni

From the figures you have presented it would appear that liquid sludge viscosity varies in the range 10 to 1000 times that of water. Considering that sludge is not a Newtonian fluid, what type of equation do you use to describe turbulent sludge flow, particularly in the context of mixing? Do you think that turbulent flow regime is necessary for good mixing of sludge in, for example, digesters?

R.C. Frost

The difference in flow behaviour between sludge and water is far more pronounced in laminar flow than it is in turbulent flow. In turbulent pipeline flow, the ration of friction head loss for sludge flow to that for water flow lies in the range 1 to 2 and increases with apparent viscosity of the sludge, having a value of about 2 at an apparent viscosity of 0.1. Head loss in turbulent flow is much less sensitive to changes in sludge rheology than is head loss in laminar flow. I am not convinced that a turbulent flow regime is necessary to achieve adequate mixing in digesters. Use of a higher solids concentration so as to avoid solids separation, may be a better alternative than increased mixing power input.

ASSESSMENT OF THE ANAEROBIC DIGESTIBILITY OF SLUDGE

D. C. STUCKEY
International Reference Centre for Waste Disposal (IRCWD)
Dubendorf, Switzerland

Summary

The importance of being able to assess, at a laboratory scale, the anaerobic digestibility of sludge is discussed together with the factors influencing this parameter. Finally, a quick inexpensive batch bioassay technique is outlined which will enable accurate assessments to be made of sludge digestibility.

1. INTRODUCTION

In recent years the processing and disposal of sludges from wastewater treatment plants is receiving increased attention. This is because sludge handling may represent as much as 30 to 40 per cent of the capital cost of a treatment plant, and about 50 per cent of the operating cost.(1) In addition, sludge disposal questions often constitute a major fraction of the problems experienced in wastewater treatment plants.

Of the methods currently available for sludge treatment and disposal, e.g. thickening and incineration, land application, aerobic digestion, anaerobic digestion, all, except anaerobic digestion, are falling out of favour in most cases due to increased chemical, energy and land costs. Hence, presently in most developed countries the method most commonly used to dispose of wastewater sludges is anaerobic digestion. This process requires no chemicals, little land, and produces a considerable amount of energy in the form of a methane rich gas (65% CH_4). In many cases the gas from a digester is sufficient to provide enough energy through dual fuel diesel engines to power a complete wastewater treatment plant.

The fraction of the organic sludge fed to a digester which is converted to gas is known as its digestibility, or biodegradability, and this fraction varies depending on the structure and composition of the sludge, and the operating parameters. It is important for design engineers and plant operators to be able to accurately assess the digestibility of sludges since this has a direct bearing on the amount of gas produced, and hence on the overall energy balance of a wastewater treatment plant. In addition, the digestibility of a sludge effects the final quantity of solids which require ultimate disposal.

In the past the anaerobic biodegradability of organic feeds has been assessed in the laboratory with small (2-10L) continuously stirred semi-continuously fed reactors. While these procedures closely simulate full-scale anaerobic digesters, they are costly in terms of facilities, equipment and personnel, and accurate data is only obtained after a considerable time has elapsed to allow for steady state. However, comparison of this laboratory data with the performance of full-scale units can provide useful insights into the efficiency of digestion. If the biodegradability of the feed sludge is considerably higher in the laboratory units than in the full-scale units then this is a strong indication that mixing in the latter units is not as effective as it should be, and that dead spots are developing leading to lower biodegradabilities. Unfortunately, since the feeds used in the laboratory and full-scale units are identical, comparisons of data will not reveal problems due to inhibition or toxicity in the full-scale unit.

In recent years due to the problems and costs involved in determining the anaerobic biodegradability of sludges with this semi-continuous assay, a batch assay has been developed which is quick, simple, inexpensive and extremely flexible. The objective of this paper is to briefly review some of the important factors controlling the anaerobic biodegradability of sludges, and then to describe a batch bioassay technique which will enable quick and accurate laboratory assessments to be made of this factor.

2. FACTORS INFLUENCING THE ANAEROBIC BIODEGRADABILITY OF SLUDGES

In wastewater treatment the two most common sludges generated are primary (raw) sludge from the primary settling basins, and waste activated sludges (WAS) generated from the aerobic treatment of soluble and colloidal organics in the settled wastewater. Primary sludge consists primarily of undigested food and dead bacteria, while WAS is mainly bacterial cells. These sludges differ in composition and structure, and hence their anaerobic biodegradabilities vary. The following discussion examines the influence of various factors on this biodegradability.

The term biodegradability is often used quite loosely in the literature. The biodegradability of complex organic materials such as sludges is difficult to measure directly. This is because when a biodegradable organic is consumed by bacteria a portion is converted to end products, such as methane, and another portion is transformed into bacterial cells. Hence, methane production, or gross organic parameters such as chemical oxygen demand (COD) reduction can only be used as indicators of the extent of biodegradation, as they give no measure of that portion converted to bacterial cells. Under anaerobic conditions the portion converted to bacterial cells may vary from 2 to 25 per cent, but under aerobic conditions it may be as high as 30 to 65 per cent. Because of this problem, operational measures such as "bioconvertability to methane" or "per cent COD destroyed" can be used only as indicators of biodegradability. However, from an engineering viewpoint these are the most relevant measures as they enable the quantity of methane produced to be assessed.

The antithesis of an organic compound which is biodegradable is one which is recalcitrant or refractory. Both have similar literary definitions; recalcitrant being stubbornly rebellious and obstinate in defying constituted authority, while refractory is defined as obstinate, intractable, unmanageable, and resisting ordinary treatment. In a biological context these terms simply indicate that the organic in question persists for extended periods in all environments tested.

In discussing what refractory compounds are, considerable confusion arises over the definition. Usually the term is defined operationally. For anaerobic digestion of organic sludges, this may mean anything that is not degraded within 35 days. However, soil microbiologists define the refractory nature of, say pesticides, in terms of years, and geochemists may think in terms of centuries. Since this paper is concerned primarily with the engineering significance of refractory compounds under normally encountered anaerobic digestion conditions, compounds not degraded after approximately 35 days will be considered as refractory. This is a purely operational definition, and will not exclude long-term biodegradability effects.

Interest in refractory organics is relatively recent, and arose with questions about the persistence of synthetic detergents and pesticides, and their concomitant effects on the environment. From this sprang a more general inquiry into the factors controlling biological degradability of both natural and synthetic organic compounds.

Until this point the question of refractory organics was dominated by the "Principle of Microbial Infallibility," which can be traced back, but is not attributed to Berjerinck (2). This principle states that either by mutation and natural selection, or by a nongenetic adaptation, one or another species appears always to develop in response to a modification in the habitat, or by the appearance of new types of carbonaceous substrates. By this adaptation it is felt by supporters of the hypothesis that every biologically synthesized organic molecule will, under some set of circumstances, be destroyed by one or more species. However, Alexander[2] has referred to the concept of microbial infallibility as one of the "grand illusions of microbiology", due to the widely documented presence of a number of naturally synthesized refractory organics. In general, the presence of refractory compounds is brought about by two causes: (a) a particular environment, or local condition may not be conducive to microbial life or to the degradation of a specific compound; and (b) the existence of substances which are resistant either totally or partially to degradation. At a more specific level, various workers have suggested a number of factors which may be important with regard to the first category, and these include; temperature (and hence microbial population); and the presence of oxygen. For the second category the considerations are more fundamental, and involve chemical structure and solubility. Each of these factors is briefly discussed below.

Under anaerobic mesophilic conditions (30-35°C), WAS is about 40 per cent degradable, while primary sludge is about 60 per cent degradable. However, at thermophilic conditions (50-60°C) there is some disagreement as to whether overall degradabilities increase. Maly and Fadrus [3] found with a mixture of primary and WAS that no significant increases in overall degradability resulted at the higher temperatures. Golueke [4] also found no significant increases in degradability with temperature. Finally, Hashimoto et al [5] found with cattle manure that the ultimate biodegradability, i.e. at infinite detention time, was independent of temperature. However, in contrast, Malina [6] found with WAS that at thermophilic temperatures volatile solids destruction was slightly higher than at mesophilic temperatures. Heukelekian [7] also found that degradability increased 25 to 30 per cent at thermophilic temperatures compared with those at mesophilic conditions. To confuse matters still further Stuckey and McCarty (8) found that for many amino acids and some proteins the mesophilic biodegradability was in fact higher than under thermophilic conditions. Based on these variable results it is difficult to conclude that temperature (and hence different microbial populations) has a consistent and predictable influence on biodegradability, and hence other factors must be more important.

In comparing the degradability of WAS under aerobic versus anaerobic conditions it is apparent that in the main there is no substantial increase, even when the growth of biomass is accounted for. Benedek (9) found that under aerobic digester conditions only 26 per cent of the volatile solids degraded in twelve days. Allowing for a yield coefficient of 40 per cent, this corresponds to an overall degradability of 44 per cent. Singh and Patterson (10) found that with a detention time of four days, volatile suspended solids destruction was 16 per cent, which represents about 27 per cent overall. However, the degradability obtainable depends on the conditions of production of the WAS. If short solids retention times are used in the activated sludge unit, then apparent degradation in aerobic digesters can be quite high because of the more easily degradable organics from the aeration tank. Finally, Stuckey and McCarty (11) found for both untreated and thermochemically pretreated WAS that the presence of oxygen did not result in significant changes in biodegradability. All these data seem to suggest that the presence of oxygen does not influence the biodegradability of sludges.

The primary causes of refractory organics seem to lie at a more fundamental level. Anaerobic digestion has not been used extensively in the treatment of industrial wastes. However, from the literature (12, 13 14, 15), it appears that high COD destruction, 75 per cent or greater, can be achieved if the substrate is predominantly soluble. This contrasts with insoluble municipal sludges and animal wastes (16), where destructions of only 40 to 60 per cent are achieved. Since the wastes treated vary widely in composition this would seem to indicate that accessibility to substrate, i.e. solubility, is one of the factors involved in the refractory nature of some organics.

Basic chemical (steric) structure also appears to have an important influence on anaerobic biodegradability. While there appears to have been little work carried out in this area, Stuckey and McCarty (8) have found for simple nitrogen compounds that comprise the building blocks of many organic nitrogen compounds, e.g. amino acids, nucleic acid bases, that simple changes in structure lead to radical changes in biodegradability. Unfortunately, this study was not able to accurately predict the biodegradability of certain simple compounds based on their structure, and when they were combined in more complex organics, e.g. WAS, this task became virtually impossible.

Since solubility and structure seem to be the most important determinants of anaerobic biodegradability, it would appear that altering these properties of an organic would affect their biodegradability. This has been found to be the case when a wide range of organics have been subjected to thermal and thermochemical treatment before digestion. This treatment increases the solubility of the organic, and the biodegradability of WAS treated at 175°C with 300 meq/L of NaOH increased by nearly 70% (11).

In summary it seems that the primary factors affecting the anaerobic digestibility of sludges are solubility,and chemical composition and structure. This would explain the differences in the biodegradability of primary sludge (≃ 60%) and WAS (≃ 40%).

3. MEASUREMENT OF ANAEROBIC BIODEGRADABILITY

As discussed earlier, in the past the most common method of assessing anaerobic biodegradability in the laboratory was with small semi-continuously fed digesters. However, in recent years, due to the drawbacks of this method a number of batch bioassay techniques have been developed.

The Warburg respirometer has been widely used as a batch procedure to evaluate biodegradability in aerobic systems, and this has been adopted for anaerobic evaluations [17, 18]. Unfortunately, this method has a number of limitations: (a) it is costly and requires some degree of skill to operate; (b) it is limited in the number of samples that can be analysed at one time; (c) sample size is limited, making subsequent analyses difficult, (d) it is difficult to sample the gas and liquid phases during the assay, and (e) extended incubation times are impractical.

In order to circumvent most of these difficulties, the procedures developed for the anaerobic Warburg were combined with serum-bottle techniques for the cultivation of anaerobes as described by Miller and Wolin [19]. Anaerobic serum bottles containing samples, defined media, and seed inocula are incubated at the desired temperature, and gas production is monitored volumetrically either by syringe [20], in a flexible bag [21] or is collected in the reactor [22]. An assay using the first method of gas measurement has been named the BMP (Biochemical Methane Potential) test, and is described in more detail below.

The BMP assay is generally carried out in 250 ml reagent bottles, although other sizes can be used. The bottles are gassed with an O_2 free mixture of 30% CO_2 and 70% N_2, stoppered with rubber serum caps, and allowed to equilibrate at the incubation temperature prior to introducing samples, defined media and inocula. This gas mixture is used since it closely approximates the percentage of CO_2 present during digestion, and results in a pH just slightly above neutrality.

The defined media is prepared from concentrated stock solutions, and contains nutrients and vitamins needed for anaerobic growth [23]. In addition, it contains sodium sulfide to provide a reducing environment, and Na HCO_3 to buffer the pH at around 7.2.

A sample of the organic being evaluated, e.g. WAS, is then transferred anaerobically to the serum bottle. Proper sample size and liquid-to-void volume ratio are important for the accuracy and precision of results, and are chosen with the following guide-lines: (a) provide a measureable, but not excessive amount of methane; (b) ensure that nutrients will not be limiting, and (c) eliminate possible substrate toxicity. Usually the estimated degradable COD is kept to less than 2g/L in the assay liquid, and the final total liquid volume is 160 ml (void volume of 104 ml).

The defined media is then equilibrated to the assay temperature (e.g. 35°C), inoculated with 10% V/V of an anaerobic seed from a seed reactor, and transferred anaerobically to the serum bottle. After the serum cap is inserted the bottles are equilibrated for one hour at the incubation temperature and then gas volumes are zeroed with a syringe and the bottles are ready for incubation and sampling. Normally, in order to obtain precise results, two or more bottles are run of the same sample together with 3-4 "seed blanks", i.e. bottles containing only media and seed.

Over a period of 30-35 days gas production is monitored periodically with a wetted glass syringe, and every time gas is wasted its composition is determined with a gas partitioner. Thus a detailed methane balance is kept for every serum bottle. At the end of the 30-35 days the volume of methane produced in the seed blank (typically 1.5 ± 0.5 ml) is subtracted from the total methane produced from degradation of the sample (typically 50.0 ± 1.0 ml), and the results expressed as m^3 CH_4 produced per kg of COD added. This allows the direct transfer of results into per cent of organic matter converted to methane, or per cent digestibility, using the theoretical relationship of $0.350m^3$ of methane (at STP) being produced for every kilogram of COD converted to methane [24].

This method is extremely precise for a bioassay technique (± 2%), is flexible since many samples can be run at one time and the test need not be terminated at 35 days, and is inexpensive to set up and operate. In addition, through a few simple modifications the assay can be used to detect toxicity.

Through the BMP assay it is possible to obtain kinetic data on the degradation of sludges, and in comparing these results with data from full-scale operation it is possible to pinpoint some problems in the full-size units due to poor mixing or short circuiting. Furthermore, due to the dilution involved in the assay, toxicity should not occur and hence any significant difference between these laboratory results and full-scale operation may indicate that inhibitory substances are present in the feed.

4. SUMMARY

The ability to be able to assess the anaerobic digestibility of sludges in the laboratory is important for design engineers and plant personnel since this data is useful in predicting the amount of energy available from an anaerobic digester, and the quantity of solids which require ultimate disposal. In addition, discrepancies between laboratory and full-scale biodegradability enables problems such as poor mixing and toxicity in the full-scale units to be quickly pinpointed and diagnosed.

While many factors have been suggested in the literature which influence the anaerobic biodegradability of sludges, it appears that the most important ones are properties of the sludge, i.e. its solubility and its chemical and physical structure. Only by altering these properties, e.g. through thermochemical pretreatment can the biodegradability of these sludges be increased.

Finally, a simple inexpensive bioassay (BMP) technique has been described which will enable the assessment of sludge digestibility under anaerobic conditions.

REFERENCES

(1) Vesilind, P.A. Treatment and Disposal of Wastewater Sludges. Ann Arbor Science Publishers, Inc., Ann Arbor, Michigan (1975).

(2) Alexander, M., "Biodegradation: Problems of Molecular Recalcitrance and Microbial Fallibility, "Adv. in Applied Microbiology, 7, 35, (1965).

(3) Maly, J., and H. Fadrus, "Influence of Temperature on Anaerobic Digestion, "J.W.P.C.F., 43, (4), 641, (1971).

(4) Golueke, C.G., "Temperature Effects on Anaerobic Digestion of Raw Sewage Sludge, "Sewage and Industrial Wastes, 30, 1225, (1958).

(5) Hashimoto, A.G., Y.R. Chen and V.H. Varel. Anaerobic Fermentation of Beef Cattle Manure - Final Report. Prepared for SERI (SERI/TR-98372-1). USDA Meat Animal Research Center, Clay Center, Nebraska (1981).

(6) Malina, J.F., Jr., "The Effect of Temperature on High-Rate Digestion of Activated Sludge, "Proceedings of the 16th Conference on Industrial Wastes, Purdue University, 232, (1961).

(7) Heukelekian, H., "Further Studies on Thermophilic Digestion of Sewage Solids, "Sewage Works Journal, 2, 219, (1930).

(8) Stuckey, D.C. and P.L. McCarty. The Effect of Thermal Pretreatment on the Anaerobic Biodegradability and Toxicity of Waste Activated Sludge. Accepted for Water Research (1983).

(9) Benedek, P., "Theory Applied to the Aerobic Stabilization of Organic Sludges Arising from Waste Water Treatment," in Design Aspects of Biological Treatment, IAWPR Short Course, Birmingham, U.K., (1974).

(10) Singh, T., and J.W. Patterson, "Improvement of the Aerobic Sludge Digestion Process Efficiency, "J.W.P.C.F., 46, (1), 100, (1974).

(11) Stuckey, D.C. and P.L. McCarty. Thermochemical Pretreatment of Nitrogenous Materials to Increase Methane Yield.Biotechnology and Bioengineering Symp. No. 8, 219-233 (1978).

(12) McCarty, P.L., D.J. Hahn, G.N. McDermott, and P.J. Weaver, "Treatability of Oily Wastewaters from Food Processing and Soap Manufacture, "Proceedings of the 27th Conference on Industrial Wastes, Purdue University, (1972).

(13) McCarty, P.L., "Methylene Chloride Toxicity and Assay in Anaerobic Digestion," Report prepared for the National Coffee Association of U.S.A., Inc., New York, (1977).

(14) Aulenbach, D.B., and R.V. Hallock, "Treatment of Milk Truck Washing Wastes by Anaerobic Digestion," Proceedings of the 27th Conference on Industrial Wastes, Purdue University, 126, (1972).

(15) Hiatt, W.C., A.D. Carr, and J.F. Andrews, "Anaerobic Digestion of Rum Distillery Wastes," Proceedings of the 28th Conference on Industrial Wastes, Purdue University, 966, (1973).

(16) Hobson, P.N., and B.G. Shaw, "The Anaerobic Digestion of Waste from an Intensive Pig Unit," Water Res., 7, 437, (1973).

(17) McCarty, P.L., J.S. Jeris and W. Murdoch. Individual Volatile Acids in Anaerobic Treatment, J.W.P.C.F, 35, 1501-1516 (1963).

(18) Kugelman, I.J. and P.L. McCarty. Cation Toxicity and Stimulation in Anaerobic Waste Treatment, J.W.P.C.F, 37, 97 (1965).

(19) Miller, T.L. and M.J. Wolin. A Serum Bottle Modification of the Hungate Technique for Cultivating Obligate Anaerobes. Appl. Microbiology, 27, 985-987, (1974).

(20) Owen, W.F., D.C. Stuckey, J.B. Healy, Jr., L.Y. Young and P.L. McCarty, "Bioassay for Monitoring Biochemical Methane Potential and Anaerobic Toxicity," Water Res., 13, (6), 485, (1979).

(21) Varel, V.H., H.R. Isaacson, and M.P. Bryant. Thermophilic Methane Production from Cattle Waste. Appl and Environ. Micro, 33, (2), 298-307 (1977).

(22) Viitasalo, 1. Ett Snabb-t Laboratorietest för Biogas-production. Biogas från godset och slam. Rapport från ett Nordiskt Seminarium. September (1980).

(23) Speece, R.E. and P.L. McCarty. Nutrient Requirements and Biological Solids Accumulation in Anaerobic Digestion. Advances in Water Pollution Research, Vol 2, p 305, Pergamon Press (1964).

(24) McCarty, P.L. Anaerobic Waste Treatment Fundamentals. Part I. Public Works, September, p 107, (1964).

ASSESSMENT OF SLUDGE STABILITY

A M Bruce
Water Research Centre
Elder Way
Stevenage, Herts, UK

Summary

The measurement of the stability of sewage sludges is one of the more difficult areas of sludge characterisation. There is no standard definition of 'stability' and no universally-applicable test to cover all the types of sludge produced by the various forms of stabilisation process.

The stability (or instability) of raw sludges can be of great importance in regard to treatment and disposal operations at sewage works. To quantify possible effects, it may be desirable to conduct laboratory tests to determine the changes which occur in certain parameters during anaerobic storage. Results of some tests are given.

The various parameters which have been used for assessing the stability of stabilised sludges are reviewed and examined. In particular, the problems associated with the use of 'volatile solids destruction' as a criterion of adequate digestion are highlighted.

INTRODUCTION

The measurement of the biochemical stability or (instability) of sewage sludges is generally regarded as one of the more difficult areas of sludge characterisation. A major problem is that there is no commonly recognised all-purpose standard parameter by which to assess sludge stability so that it usually becomes necessary to adopt different tests for different purposes and for different types of sludge. The development of a single standard stability test was an early objective of the COST 68 work[(1)] and the search is still continuing though without any immediate prospect of success.

The practical interest in, and importance of, assessing sludge stability relates to two different aspects of sludge treatment and disposal. The first area of interest concerns the stability of raw sludges because the rate and degree of change in certain properties of such sludges during even short-term anaerobic storage may have a profound influence on the design, operation, and performance of subsequent processing stages. The second, and major, area of interest in stability measurement relates to the need to assess the stability of stabilised sludges so that the performance of the stabilisation process itself can be evaluated and monitored and the nuisance potential of the sludge during disposal can be quantified.

STABILITY OF RAW SLUDGES

Raw primary sludges, and raw secondary sludges from biological treatment plants operating at high or medium organic (F/M) loadings, are biochemically very active and may exhibit rapid and deleterious changes in their properties when stored anaerobically. The changes which are of main practical importance, and which occur as a result of anaerobic biodegradation (involving solubilisation of organic solids, volatile acid production and cell lysis) are increases in the following:-

(a) Oxygen demand of the sludge liquor

(b) Specific resistance to filtration and coagulant demand

(c) Concentration of non-settleable solids (fines)

(d) Offensiveness and intensity of odour

The potentially adverse effects which these changes may have on other treatment and disposal operations are indicated in Table 1.

Table 1. Potentially adverse effects of anaerobic storage of unstable raw sludges on treatment and disposal operations

Storage causes increase in	Potential adverse effect on treatment/disposal operation			
	Gravity thickening	Mechanical dewatering	Direct disposal to land	Treatment of sludge liquors
Odour intensity	Odour nuisance	Odour nuisance	Odour nuisance	-
BOD/COD of sludge liquor	-	-	-	Increased load on secondary plant
Non-settleable solids	Reduced solids capture	Reduced solids capture	-	Increased load on secondary plant
Specific resistance to filtration and coagulant demand	-	Increased chemical costs	-	-

The rate and extent to which the changes occur during storage, however, vary considerably with the type of sludge and there is often a need to carry out laboratory stability tests on a particular sludge in order to quantify potential effects. The tests developed at the Water Research Centre involve anaerobic storage of a sample of the sludge in a WRC standard settling apparatus(2) (Fig. 1) at a constant temperature (usually 20°C though any appropriate temperature may be selected). The sludge is stirred at 1 rev/min. Samples of the sludge are withdrawn at appropriate intervals i.e. hourly, or daily, depending on the degree of instability of the sludge, and subjected to all or some of the following tests:-

(a) BOD and COD of liquid phase

A portion of the sludge is filtered through a GF/C paper and standard determinations of BOD_5 and/or COD are made on the filtrate. Where the filtration rate of the sludge is too slow, centrifugation may be employed to obtain a sample of sludge liquor.

(b) Filtrability and coagulant demand

The CST or specific resistance to filtration of the sample is determined and its conditionability assessed by the standard method[1].

(c) Non-settleable solids

A sample of the sludge is diluted 1:1 (V/V) with water and allowed to settle in a measuring cylinder for 24 hours at a constant temperature (usually 20°C). A sample of the supernatant is then removed and the concentration of suspended solids determined by standard method.

(d) Offensiveness and intensity of odour

This may be assessed by a simple qualitative 'nose' test by an odour panel but, if a more quantitative assessment is required, a specific volume of odour-free air can be passed through the sludge and the odour intensity of the exit air measured using a standard dynamic dilution technique[3]. The exit air can also be used for analysis of the volatile compounds in the sludge using chemical or gas chromatographic methods although interpretation of results may be difficult.

Results from (a) and (c) are usefully expressed as mg/g sludge solids or as a rate of change i.e. mg/g.h over the period of the test. It is clear from Table 1 that not all of the tests need to be done in all cases.

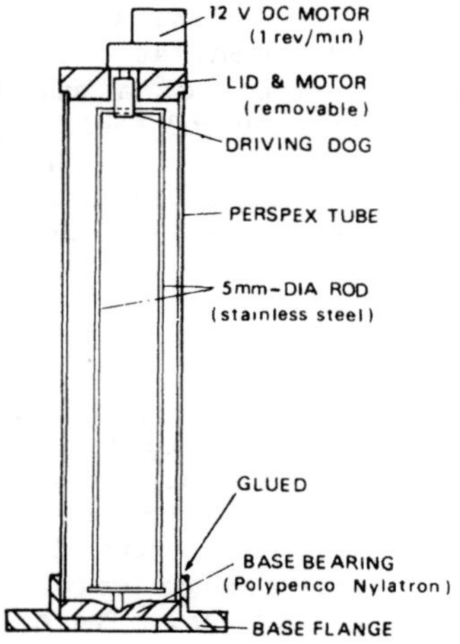

FIG. 1. Standard settling apparatus

Some representative results from tests are given in Tables 2 and 3. Table 2 shows the the greater instability of sludge from a highly-loaded activated sludge plant compared to that of sludge from a low-rate plant. Table 3 demonstrates the extreme instability of a humus sludge from a high rate biological filter.

White et al[4] have shown that storage for 2-3 days of a high-rate activated sludge resulted in more than a doubling of the cost of conditioning the sludge for mechanical dewatering. This factor would obviously need to be taken into account in assessing the overall cost effectiveness of gravity thickening before mechanical dewatering. In other tests, it has been shown that some primary sludges remain relatively stable over several days after showing an initial rapid deterioration in characteristics during the first few hours in the settlement tank. In such cases, the subsequent period of anaerobic storage in, say, a gravity thickener would not have any significant adverse effect on sludge quality.

Such results point to the need for quantitative information on the stability of raw sludges in certain cases and the tests indicated above may usefully serve this purpose.

STABILITY OF STABILISED SLUDGES

A good deal of attention has been given over the years to the development of methods for assessment of the stability of stabilised sludges but, as indicated earlier, there is as yet no standard test applicable to all types of stabilisation process and all types of sludge. Vesilind[5] comments that 'sludge stabilisation is often practiced without any consideration of what is meant by a stable sludge and how and why this must be obtained'. There is a move now, however, towards the adoption of specific criteria of satisfactory stabilisation in national guidelines or regulations relating to the disposal of sludge to land even though the criteria used may be recognised as deficient in some respects.

It can be argued that if the main objective of stabilisation is odour control then the basis for a standard stability test should be the odour characteristics of the sludge. This approach has been taken in some cases[6] but odour assessment is complex and is unsuitable as a routine quality control test at, say, a sewage works laboratory. A number of other parameters based on physical chemical, or biological properties have been proposed, or are actually used, as a means of assessing stabilised sludge quality or the performance of a stabilisation plant. Table 4 gives a summary of these parameters and their applicability to the various methods of stabilisation. It is seen that most of the tests are applicable only to liquid sludges from anaerobic, aerobic or lime stabilisation processes and that there are few tests relevant to composted solid sludges.

A. ODOUR CHARACTERISTICS

It is generally recognised that the compounds that can cause odour nuisance from sludges include hydrogen sulphide, skatole, indole mercaptans, organic sulphides, volatile fatty acids and amines. However, detection by specific tests of the presence of one or more of these compounds would not necessarily mean that the sludge was unstable (i.e. odorous) and conversely their absence would be no guarantee that the sludge would not cause odour nuisance. There is, for example, a general

Table 2. Changes during storage at 20°C in the characteristics of surplus activated sludge from a low-rate* plant (A) and a high-rate* plant (B)

Period of storage at 20°C (h)	Specific resistance to filtration (10^{12} m/kg at 49 kPa)		Non-settleable solids (mg/l)		COD of filtrate (mg/l)	
	A	B	A	B	A	B
0	19	3216	4	56	48	160
1	-	-	4	124	42	205
3	32	4335	-	-	-	-
14	-	-	28	484	64	265
24	183	>10000	56	448	76	550

* Average organic loading on Plant A = 0.1 kg BOD/kg MLSS d

" " " " " B = 3.0 " " " "

Table 3. Changes in BOD of filtrate from a high-rate humus sludge during storage at 20°C
(Initial solids content of sludge = 2.1 per cent.)

Period of storage (d)	BOD of filtrate mg/l	mg/g solids
0	700	33.3
1	6000	286
2	9500	452
3	10000	476
4	13000	619

Table 4. Possible methods for measurement of the stability of sewage sludges from stabilisation processes

Basic parameter	Measurement	Possible applicability to each method of stabilisation				Comments
		Anaerobic digestion	Aerobic digestion	Lime stabilisation	Composting	
Odour emitted by the sludge	1. Odour intensity by dilution technique 'Threshold dilution value'	√	√	√	√	Requires an odour panel. Not related to odour 'quality'
	2. Gas-chromatography/mass spectrometry	√	√	√	√	Expensive and difficult to interpret
	3. Attractivity to house flies	√	√	√	√	Tentative. Further tests required
Volatile solids	4. Volatile solids in the sludge as fraction of the total solids	√	√	x	?	Standard measurement (see text)
	5. Fraction of volatile solids destroyed (FVSD) by the stabilisation process	√	√	x	?	Standard measurement (see text)
Residual, readily-biodegradable matter	6. BOD_5 of filtrate	x	√	x	x	Standard measurement
	7. Rate of increase of BOD or COD of filtrate with storage	√	√	x	x	Tentative basis for a 'stability index'
	8. Specific oxygen uptake rate (SOUR)	x	√	x	x	Temperature dependent

Table 4. (Cont'd)

Basic parameter	Measurement	Possible applicability to each method of stabilisation				Comments
		Anaerobic digestion	Aerobic digestion	Lime stabilisation	Composting	
	9. Gas production during anaerobic incubation at 35°C	√	√	x	x	Standard control sludge required to detect inhibition
Chemical composition	10. Volatile fatty acids	√	√	x	x	Standard analysis
	11. pH and pH change during storage	√	√	√	x	Standard test
	12. H_2S emission on storage	x	√	x	x	Special test
	13. Nitrate concentration	x	√	x	x	Standard analysis
	14. Grease	√	√	x	x	Standard analysis
Biological activity	15. ATP, DNA concentration	x	√	x	x	Research application only
	16. Enzyme concentration	x	√	x	x	Research application only

√ Applicable

x Non-applicable

strong correlation between volatile fatty acid concentration and the offensiveness of a sludge[7] but the influence of pH is very marked and there are also synergistic, counteractive, and masking effects which make predictions about odour characteristics from chemical analyses alone very unreliable.

The human nose is obviously a very sensitive detector of odour and is perhaps the most appropriate 'instrument' to use for assessing the quality and intensity of odour from a sludge. In some cases, subjective assessments of the quality of an odour can be useful and for more quantitative evaluations of 'odour intensity' or 'odour number' dilution techniques may be employed using an odour panel to assess the degree of dilution with odour-free air that is required before the odour of a sample of sludge, or odorous air from it, is just not detectable.

Instrumental methods for analysing odorous air, using gas-chromatography/mass-spectroscopy techniques are now well developed. They can provide a mass of detailed information about the volatile compounds present in the sludge. However, as with chemical analyses, the information cannot be readily interpreted to assess 'offensiveness' or 'stability'. For the time being, such techniques remain suitable only for research purposes. It is conceivable that in the future an automatic computer-based' odour analyser will be developed for routine testing of sludge stability.

Studies in the USA[8] have shown that the attractivity of sewage sludge to house flies decreased with the degree of stability of the sludge. The possibility of using this type of 'bio-assay' test as a means of characterising the odour quality of a sludge (and its potential to cause nuisance) may be worth further investigation.

B. VOLATILE SOLIDS

The determination of total sludge solids (TS) and sludge volatile solids (VS) as %W/W is a routine test for most sewage works laboratories. For sludges that are anaerobically or aerobically digested, the ratio VS/TS (loss on ignition) reduces from about 0.75 for the raw sludge to about 0.62 for the digested sludge owing to the destruction of about 40% of the volatile matter. At a given works, it is possible to establish a normal range of values of VS/TS for well-digested sludge and any values above the range could indicate a poorly-digested and unstable sludge. The main problem with VS/TS as a parameter of stability is that changes in the composition in the raw sludge (e.g. an increase in the proportion of secondary sludge) could affect the normal ratio of VS/TS for a well digested sludge. Normal, day-to day, variations of VS/TS could also mask a small degree of contamination of the digested sludge with raw sludge. Thus the ratio VS/TS is not a sufficiently reliable parameter to be used as a general criterion of sludge stability. Furthermore, it is not applicable to some chemical stabilisation techniques such as lime stabilisation where there is no reduction in VS on treatment.

C. FRACTION OF VOLATILE SOLIDS DESTROYED

The fraction (expressed as a percentage) of raw sludge volatile solids destroyed by an anaerobic or aerobic digestion process has been used traditionally as a performance parameter and as an indicator of the degree of digestion (stability) of the digested sludge. It is also now being used as a more formal criterion of adequate digestion. In the UK national guidelines[9] on disposal of sludge to land, the achievement of 40 per cent reduction in volatile solids by digestion is regarded as evidence of 'effective digestion'. Values of less than 40 per cent imply inadequate digestion and a rather more restrictive use of the sludge on pasture land is required. In the case of the US EPA regulations[10], the minimum requirement for volatile solids destruction is 38 per cent for both aerobic and anaerobic digestion.

The general experience is that in the anaerobic digestion of raw (co-settled) sewage sludge sludges, the fraction of volatile solids destroyed is usually between 40 and 50 per cent across the primary digester. The percentage destroyed across the primary and secondary digester together could be somewhat greater. In view of this, the requirements expressed above seem reasonable. However, lower reductions in volatile matter than 40 per cent would normally occur if the sludge being digested was mainly secondary sludge or if partial digestion of the raw sludge had occurred prior to entering the primary digester, though in both cases the sludge might still be fully digested. With aerobic digestion, the maximum reduction in volatile matter is often less than 40 per cent even with complete digestion.

There are some other basic problems in using the fraction of volatile solids destroyed as a criterion of a stabilised sludge. These have been fully described in another paper[11] and will only be briefly outlined here:-

(a) Method of calculation of fraction of volatile solids destroyed (FVSD)

Essentially, there are two common methods for calculation of FVSD which can give significantly different results. The methods are

(i) van Kleeck's formula (ΔVK)

$$\text{FVSD} = 1 - \left[\frac{wv/wv{+}wnv \quad x \quad Wnv/Wv{+}Wnv}{Wv/Wv{+}Wnv \quad x \quad wnv/wv{+}wnv}\right]$$

and (ii) approximate mass balance formula (ΔMB)

$$\text{FVSD} = 1 - \frac{wv}{Wv}$$

where Wv and Wnv = volatile solids (%) and non volatile solids (%) respectively in the raw sludge, and

wv and wnv = volatile solids (%) and non-volatile solids (%) respectively in the digested sludge

In the ideal situation, where there is no deposition of non-volatile matter (e.g. grit) in the digester or accumulation of organic solids as scum, the two methods of calculation should give identical results but there are conditions when the calculations produce significantly different results from the same basic data. If, for example, a proportion of the non-volatile matter in the raw sludge continually becomes deposited in a digester, then calculation by van Kleeck's method will give lower values of FVSD than those obtained by the approximate mass balance method of calculation. Thus, if 5 per cent of the the total non-volatile matter was being deposited as grit, a true value of FVSD of 41 per cent would become 37.9 per cent by van Kleeck's method and 40.3 per cent by the mass balance method. Therefore both methods would give a strictly incorrect value but the former method would indicate "unsatisfactory" digestion and the latter method "satisfactory" digestion for the same situation.

In practice, it appears that van Kleeck's method is most widely used but the mass balance method has much to recommend it.

(b) Insensitivity to contamination by undigested sludge

The value of FVSD may give a reasonable indication of the degree of digestion achieved but it does not provide a reliable indication of the actual quality of the digested sludge. For example suppose that:-

(i) a digester is providing complete digestion to 95 per cent of the total quantity of raw sludge fed to it.

(ii) the remaining 5 per cent of sludge short-circuits the digester and is discharged with the digested sludge, and

(iii) the FVSD for the portion of sludge which is fully digested is 50 per cent (a not unusual value)

then the overall FVSD for the contaminated sludge discharged from the digester would be 48.4 per cent. This value is still, ostensibly, satisfactory (well over 40 per cent) yet the product would be contaminated with 5 per cent of raw sludge. This degree of contamination might not be important in terms of odour but could represent an enhanced risk of pathogen contamination.

D. GAS PRODUCTION DURING ANAEROBIC INCUBATION

Measurement of gas production over a 24 hour period during anaerobic incubation at 35°C in the laboratory can give a much more direct and sensitive indication of sludge quality than can FVSD calculations in terms of the presence or absence of undigested sludge. Thus, if a norm for the residual specific gas production rate from a well-digested sludge at a given works is determined (as litres gas/grams volatile solids remaining in the digested sludge), any significant increase in gas production rate would indicate the presence of undigested sludge. A control is necessary in all cases to ensure that other factors (e.g. inhibition) are not influencing the rate of production. The principle can also be applied to aerobically-digested sludge but an actively digesting anaerobic sludge is then required as a seed.

E. VOLATILE FATTY ACIDS

The presence of volatile fatty acids (VFAs) in high concentrations is indicative of an unstabilised sludge and odour intensity has been correlated with VFA concentration[12]. The actual odour quality of a sludge will depend on its pH value; lime-stabilised sludges may have high concentrations of VFAs without necessarily causing odour problems. Conversely, the absence of VFAs would not necessarily mean that the sludge was stable or non-odorous. A test for VFAs is not sufficient in itself to measure stability.

F. OXYGEN DEMAND

As a measure of stability, the oxygen demand of a sludge may be an appropriate test to employ for aerobically digested sludges. It has been found[6] that the specific oxygen uptake rate (SOUR) of a sludge, expressed as mg O_2/g vs h can be correlated with a subjective assessment of odour intensity. Values of less than about 2 mg O_2/g vs h at 18°C are indicative of a stable sludge, but the threshold value is strongly dependent on temperature and no general criterion can be proposed. Measurement of SOUR is not a useful method for anaerobically-digested sludges.

As indicated earlier, the rate of change in the COD or BOD of the sludge liquor at a temperature of 20°C may give an indication of relative stability. It has been found at the Water Research Centre[13] that rates of increase of COD of 1 mg/g TS.h or less are indicative of very stable surplus activated sludges. Values of 30 mg/g.TS.h were found with very unstable and malodorous sludges.

Recent work with piggery slurry, at another institute,[12] has shown a very strong correlation between the BOD of slurry liquors and the "offensiveness of odour" of the slurry as judged by an odour panel using an arbitrary scale of 'offensiveness'.

G. NITRATE

The presence of nitrate in an aerobically-digested sludge after storage is a definite indication of stability[6]. However, the occurence of flotation would indicate denitrification and incomplete stabilisation. The absence of nitrate from a sludge is not necessarily an indication of instability since it would not be present in anaerobically-digested sludges.

H. pH VALUE

Anaerobically-digested sludges are well buffered and pH may not be significantly reduced by the presence of undigested material or raw sludge. Values of pH below 7 are, however, definitely indicative of instability for sludges from all types of stabilisation process. It has been recommended[10] that the pH value of a sludge from a lime stabilisation process should remain "above 11.0 during 14 days of storage in an open beaker". It is doubtful, however, that many lime stabilisation plants operate to this stringent criterion.

I. EVOLUTION OF HYDROGEN SULPHIDE

Hydrogen sulphide is the most easily detected malodorous compound associated with unstable sludges. A method of measuring the "stability index" of a sludge, based on its potential for generating hydrogen sulphide, has been proposed[14]. Samples of sludge are incubated at 37°C in stoppered tubes containing lead acetate paper. The time taken for the paper to blacken is the stability index. The method is applicable only to aerobically digested sludges.

J. BIOCHEMICAL ACTIVITY

Quantitative indicators of biochemical activity, such as concentration of ATP or DNA or concentrations of hydrolytic or proteolytic enzymes have been used as parameters for measurement of sludge stability and a general stability test based on such measurements has been proposed by Colin[15]. The higher the level of biochemical activity, the greater the degree of sludge instability. However, the complexity and cost of the biochemical tests, and the difficulty in interpretation of the results restricts the use of such methods to the research field.

K. GREASE

Mosey[16] has suggested recently that the concentration of grease in a sludge gives an indication of completeness of digestion and this approach may have value in some cases for assessing stability. The disadvantage is that the determination of grease, by solvent extraction techniques, is rather time-consuming. In other cases, well-stabilised sludges have been found to have high "grease" contents due to the presence of non-offensive mineral oils.

CONCLUSIONS

1. There is no universally-applicable test to assess the stability of sewage sludges. It seems unlikely that one will be found.

2. The stability (or instability) of raw sludges can have a significant effect on other treatment and disposal operations at works. Tests to assess the effect of anaerobic storage on raw sludges have been employed to quantify possible adverse consequences on particular treatment operations.

3. Odour control is the main objective of stabilisation yet odour measurement is not suitable as a routine monitoring procedure for assessing sludge stability.

4. The use of 'fraction of volatile solids destroyed' as a measure of satisfactory digestion is incorporated in some national criteria. However, interpretation of the results of calculation of FVSD must be treated with caution in some circumstances.

5. Other methods of assessing sludge stability are available but in the main they are applicable only to particular forms of stabilisation and some are suitable only for research purposes.

6. At present, both works operators and those in the research field, remain in need of basic standard tests which they can use to assess sludge stability without difficulty in interpretation. There is still an incentive therefore for further development work in this area of sludge characterisation.

REFERENCES

1. COMMISSION OF THE EUROPEAN COMMUNITIES. COST Project 68. Sewage Sludge Processing. Final Report of the Management Committee. 1975.
2. WHITE, M.J.D. Design and control of secondary settlement tanks. Water Pollution Control. 1976, 459-467.
3. BEDBOROUGH, D.R. In 'Odour Control - A Concise Guide'. Ed. F.H.H. Valentin and A.A. North. Stevenage, Warren Spring Laboratory 1980, pp 17-29.
4. WHITE, M.J.D., BASKERVILLE, R.C., and LOCKYEAR, C.F. Continuous thickening of biological sludges and the influence of stability. Water Pollution Control, 1977, 86-97.
5. VESILIND, P.A. Treatment and Disposal of Wastewater Sludges. Ann Arbor Science, 1975.
6. EIKUM, A.S. and PAULSRUD, B. Methods for measuring the degree of stability of aerobic stabilised sludges. Water Research, 1977, 11, 763-770.
7. LEADBEATER, B., DIAPER, J., TAYLOR, and MELLORS, S. Odours from sewage sludge - Interim report to Water Research Centre, University of Bradford, 1981.
8. WOLF, H.W. Housefly breeding in sewage sludge. Sewage and Industrial Wastes, 1955, 27, 2, 172-176.
9. STANDING COMMITTEE ON THE DISPOSAL OF SEWAGE SLUDGE. Report of the Sub-Committee on the Disposal of Sewage Sludge to Land. Standing Technical Committee Report No 20. Department of the Environment and National Water Council. London, 1981.
10. US ENVIRONMENTAL PROTECTION AGENCY. Federal Register 1979, 44, No 179. pp53438-53464.
11. FISHER, W.J. Calculation of volatile solids destruction during sludge digestion. Support paper at the WRC Conference on Stabilisation and Disinfection of Sewage Sludge. Manchester, April 1983.
12. WILLIAMS, A.G. Measurement and control of piggery slurry odours. Poster paper presented to the WRC Conference on Stabilisation and Disinfection of Sewage Sludge. Manchester, April 1983.
13. BASKERVILLE, R.C. Unpublished work at the Water Research Centre, UK.
14. HARTMAN, R.B., BENNETT, E.R. and LINSTEDT, K.D. New procedure determines sludge stability. Water and Sewage Works. 1978, 125, 4, 42-44.
15. COLIN, F. Methodes d'evaluation de la stabilite biologique des boues residuaires. Report No WPSP28. COST 68 Working Party No 1. Commission of the European Communities 1980.
16. MOSEY, F.E. and FOULKES, M. Control of the anaerobic digestion process. Paper presented to the WRC Conference on Stabilisation and Disinfection of Sewage Sludge, Manchester, April 1983.

DISCUSSION

P. Balmer

Because of the problems associated with estimation of the fraction of volatile solids destroyed, we might consider using gas production as an alternative measure of solids conversion. The input organic matter could be measured directly as COD, while the removed COD could be calculated from the gas production rate.

A.M. Bruce

I would agree with that suggestion. There may, however, be a practical problem in that currently used systems for gas measurement may not be very reliable. Prof. Balmer has also proposed another method, namely, differential thermal analysis, but I think this must still be regarded as a research rather than as an operational control parameter.

CONCLUSIONS

I. Thickening and Dewatering of sludges

B. Paulsrud

Several parameters for the characterisation of sludge for the thickening and dewatering processes have been discussed in this workshop. The table below is an attempt to summarise these characterisation methods with a view to evaluating their usefulness in prediction of full scale performance or process design or process control

CHARACTERISATION METHOD	GRAVITY THICKENING		MECHANICAL DEWATERING		
(PARAMETER)	Batch	Continuous	Centrifuge	Belt Press	Filter Press
Frozen image centrifuge test	)				
Pilot scale thickener	)+				
Specific resistance to filtration					+
Capillary suction time (SCT)					+
Coagulant demand (standard conditioning test)				))+	
Modified filtration test				)	
Rheology (viscometer test)				?	
Penetrometer test			-		
Laboratory scale centrifugation			-		
Floc strength (standard stirrer test)			?		

+ = Reliable for prediction of full scale performance/process design/process control.

From this table it is evident that there is still a need for research on methods for assessing continuous gravity thickening and dewatering by centrifuges. More work should also be done on the use of rheology, especially for belt presses, where the viscometer test is very promising.

II. Sludge Stabilisation: Physical State of Sludge

P.H. Brunner

Characterisation of sewage sludge is needed in order to design and operate sludge treatment processes properly. While for some parameters methods of characterisation have been developed and approved, others are still being evaluated or even looked for.

This workshop gave an overview on the current knowledge of methods to assess stabilisation, dewatering and handling of sewage sludge. It became obvious, that despite large efforts to find laboratory experiments to predict or simulate actual field conditions there lacks information to characterise some very basic sludge treatment processes. The reason for this is quite often not the treatment process itself but the sewage sludge, which still is a rather ill defined substance from a physical-chemical point of view.

One parameter which needs further characterisation is sludge stabilisation. It is interesting to note that sludge stabilisation is performed by various means in very many full scale plants but there exists not standard definition for the exact goal of this process. There are many methods available to describe some of the aspects of stabilisation, but none of them has been developed to the point of a standard test of the whole process. Odour control, being the main objective of stabilisation will be restricted to research purposes only unless individual trace substances can be singled out. Since most other parameters like volatile solids, gas, fatty acids pH etc. have been investigated throughly in the past, it might seem necessary to look for entirely new concepts, e.g. bioassay techniques, to characterise stabilisation. Otherwise it could well be that we have to face the fact that there is no single appropriate method to characterise stabilisation.

Various approved methods do exist for the assessment of anaerobic sludge digestibility. The question arises which of the available methods is best and most effective to be applied to a specific problem. From the discrepancy between the laboratory assessment test and actual field performance malfunctioning of digesters can easily be detected. In any case the alteration of sludge properties by physical-chemical pretreatment has to be taken into account. It does not seem to be necessary to put additional effort into the assessment of anaerobic digestibility of sewage sludge.

As mentioned, sewage sludge is not a well defined substance. It is therefore most often difficult if not impossible to explain sludge characteristics, e.g. rheological parameters, by the physical-chemical composition of a sludge. Nevertheless, progress has been made to compare various microscopic physical properties of sludges even if the underlying reasons, e.g. physical-chemical sludge particle characterisation, are not known. By the physical characterisation of a sludge according to rheological parameters it becomes possible to improve the design and operation of sludge handling (storage, transport, etc.). Attempts are going as far as to design entire sludge treatment schemes according to the physical properties of sludges. It seems evident that the research into rheology of sludges has reached a very rewarding and fruitful stage.

LIST OF PARTICIPANTS

ANDERSON, J
Dept. of the Environment
O'Connell Bridge House
IRL-DUBLIN 1

AVERILL, I
Dublin Corporation
28, Castle St.
IRL-DUBLIN 2

P. BALMER
Chalmers University of Technology
Water Supply and Sewerage Engineering
Fack
S-41296 GÖTEBORG

BERG, N
Norwegian Institute for Water Research
P.O. Box 333,
Blindern
N-OSLO 3

BRADLEY, P
Harper & Fay Ltd.
Beechill Clonskeagh
IRL-DUBLIN 4

BRADLEY, T
Donegal Conty Council Offices
Lifford
IRL- Co. DONEGAL

BRADY, P
Pressure Cleaners Ltd. Unit T5
Stillorgan Industrial Park
IRL- Co. DUBLIN

A. BRUCE
Water Research Centre
Elder Way
Stevenage
UK-HERTS SG1 1TH

P.H. BRUNNER
EAWAG
CH-8600 DUBENDORF

CALCUTT, T
Water Research Centre
Stevenage Laboratory
Elder Way
Stevenage
UK-HERTS SG1 1TH

CAMPBELL, H.W.
Waste Water Technology Centre
Environmental Protection
Service Environment Canada
P.O. Box 5050
CAN-BURLINGTON ONTARIO L7R 4AE

CARLEY, Mr.
Wexford County Council
Country Hall
IRL-Co. WEXFORD

CASEY, T.J.
University College Dublin
Department of Civil Engineering
Earlsfort Terrace
IRL-DUBLIN 2

CASSIDY, P
T.J. O'Connor & Associates
Belfield Goatstown Rd.
IRL-DUBLIN 14

CAWLEY, Mr.
Dept. of the Environment
O'Connell Bridge House
IRL-DUBLIN 1

CLARKE, P
Kildare County Council Offices
Naas
IRL-Co. KILDARE

CLEARY, F
County Council Offices
IRL-CAVAN

COLIN, F
Institut de Recherches
Hydrologiques de Nancy
10, rue Ernest Bichat
F-54000 NANCY

COLLINS, J
Dublin Corporation
28, Castle St.
IRL-DUBLIN 2

CONNOR, B
Institute for Industrial Research
& Standards
Industrial Est 11
IRL-DUBLIN

COUGHLAN, F
County Council Headquarters
John's Gn.
IRL-KILKENNY

COUGHLAN
Cork County Council County Hall
IRL-CORK

CREED, E.J.
Nicholas O'Dwyer Son &
Partners
Dundrum Centre
IRL-DUBLIN 14

DALY, J
Dept. of the Environment
O'Connell Bridge House
IRL-DUBLIN 1

EVANS, J
Dept. of the Environment
O'Connell Bridge House
IRL-DUBLIN 1

EVESON, J
Dept. of the Environment
Stormont
BELFAST - NTH. IRELAND

FITZPATRICK, M
J.B. Barry & Partners
Tramway Hse.
43, Dartry Rd.
IRL-DUBLIN 6

FROST, R
Water Research Centre
Stevenage Laboratory
Elder Way
Stevenage
UK-HERTS SG1 1TH

FULHAM, P
Dublin Corporation
28, Castle St.
IRL-DUBLIN 2

GALLAGHER, P
County Council Offices
IRL-CAVAN

GIBBONS, F
County Council Offices
IRL-CAVAN

GRANAHAN, J
Mayo County Council
Castlebar
IRL-Co. MAYO

HAUGH, A
T.J. O'Connor & Associates
Belfield Goatstown Rd.
IRL-DUBLIN 14

HEIDE, B.A
TNO
P.O. Box 214
NL-2600 AE DELFT

HIGGINSON, J
County Council Offices
IRL-CARLOW

HOARE, E
Wexford County Council
County Hall
IRL-Co. WEXFORD

HOYLAND, G
Water Research Centre
Stevenage Labortory
Elder Way
Stevenage
UK-HERTS SG1 1TH

KARPER, R
Rijksinstituut voor de Zuivering
van Afvalwater
Maerland 5/6
Postbus 17
NL-8200 AA LELYSTAD

KOUZELI-KATSIRI, A
National Technical University
Chair of Sanitary Engineering
5, Iroon Polytechniou
St. Zografou
GR-ATHENS 624

KELLAGHAN, P.J.
County Council Offices
Navan
IRL-Co. MEATH.

KERR, H
County Council Offices
County Hse.
Lifford
IRL-Co. DONEGAL

KILLEEN, R
Galway County Council
County Buildings
Prospect Hill
IRL-GALWAY

LAWTON, T
E.G. Pettit & Company
39, Leeson Pk.
IRL-DUBLIN 6

LAVELLE, M
Wexford County Council
County Hall
IRL-Co. WEXFORD

LESCHBER, R
Institut für Wasser-, Boden- und
Lufthygiene des Bundesgesundheitsamt
Corrensplatz 1
D-1000 BERLIN 33

L'HERMITE, P
DG XII/G-1
Environmental Protection and
Climatology
Rue de la Loi 200
B-1049 BRUSSELS

LINNANE, M
Dublin Corporation
28, Castle St.
IRL-DUBLIN 2

LOLL, U
Abwasser-Abfall-Aquatechnik
Heinrichstr. 10
D-6100 DARMSTADT

LOOBY, M
Offaly County Council
Court Hse.
Tullamore
IRL-Co. OFFALY

LUCEY, P
Cork County Council
County Hall
IRL-Co. CORK

MANSFIELD, E.J.
Waterford County Council
Arus Brugha
Dungarvan
IRL-Co. WATERFORD

McGOVERN, P.J.
Nicholas O'Dwyer Son & Partners
Dundrum Centre
IRL-DUBLIN 14

McKENNA, J
Harper & Fay Ltd.
Beechill
Clonskeagh
IRL-DUBLIN 4

McMAHON, E
Institute for Industrial Research
and Standards
Industrial Est. 11
IRL-DUBLIN

McMAHON, N
P.J. Sheahan & Partners
St Munchin's hse
IRL-LIMERICK

McMANUS, J
County Council Offices
IRL-CAVAN

MININNI, G
Istituto Ricerca sulle Acque
Università di Roma
via Reno 1
I-00198 ROMA

MONAHAN, F
Galway County Council
Co. Buildings
Prospect Hill
IRL-GALWAY

MULVIHIL, P
Westmeath County Council
County Buildings
Mullingar
IRL-Co. WESTMEATH

MURTAGH, P
Longford County Council
Church St.
IRL-Co. LONGFORD

NEWELL, P.J.
Dept. of Civil Engineering
University College Galway
IRL-GALWAY

NEWMAN, P.J.
Water Research Centre
9 Darmouth Street
UK-LONDON SW1H 9BL

O'BRIEN, B
Ferry House
Lower Mount St.
IRL-DUBLIN 2

OCKIER, P
Zandvoordestraat 375
B-8400 OOSTENDE

O'CONNELL, A
Limerick county Council
HQ County Buildings
O'Connell St.
IRL-LIMERICK

O'CONNOR, P
Dept. of Civil Engineering
University College
Earlsfort Terrace
IRL-DUBLIN 2

O'DONOGHUE, P
Kerry County Council
County Hall
Tralee
IRL-Co. KERRY

O'FLYNN, J
Tipperary (N.R.) Co. Council
Court Hse.
Nenagh
IRL-Co. TIPPERARY

O'KEEFFE, M
Tipperary (N.R.) Co. Council
Court Hse.
Nenagh
IRL-Co. TIPPERARY

O'SULLIVAN, D.P.
Delap & Waller
63 Lr. Baggot St.
IRL-DUBLIN 2

PATTERSON, B
Irish Specialists Treatment
Systems Ltd.
Greenvale Mill
Ardee
IRL-Co. LOUTH

PAULSRUD, B
Norwegian Institute for Water Researc
P.O. Box 333
Blindern
N-OSLO 3

PHELAN, P
Leitrum County Council
Court House
Carrick-on-Shannon
IRL-Co. LEITRUM

PUOLANNE, Y.J.
National Board of Waters
Pohjoinen Rautatiekatu 21B
P.O. Box 250
SF-00101 HELSINKI 10

REILLY, J.V.
Mayo County Council
Castlebar
IRL-Co. MAYO

ROBINSON, C
P.J. Sheahan & Partners
St Munchin's Hse
IRL-LIMERICK

RUSSELL, O.A.
Delap & Waller
63 Lr. Baggot St.
IRL-DUBLIN 2

SCHELTINGA, H.M.J.
Staatstoezicht op de Volksgezondheid
Hoofdinspectie voor de Hygiene van
het Milieu
Pels Rijckenstraat 1
NL-6800 DR ARNHEM

SHAW, P.J.
Westmeath County Council
County Buildings
Mullingar
IRL-Co. WESTMEATH

SPINOSA, L.
IRSA
Riparto Sperimentale di Bari
Via Francesco de Blasio 5
I-70123 BARI

STUCKEY, D
EAWAG
CH-8600 DUEBENDORF

TORPEY, P
Pressure Cleaners Ltd.
Unit T5
Stillorgan Industrial Park
Stillorgan
IRL-Co. DUBLIN

WALDRON, D
Mayo County Council
Castlebar
IRL-Co. MAYO

WALSH, P
Dublin County Council
46 Upr. O'Connell St.
IRL-DUBLIN 1

INDEX OF AUTHORS

GPSR Compliance

The European Union's (EU) General Product Safety Regulation (GPSR) is a set of rules that requires consumer products to be safe and our obligations to ensure this.

If you have any concerns about our products, you can contact us on ProductSafety@springernature.com

In case Publisher is established outside the EU, the EU authorized representative is:

Springer Nature Customer Service Center GmbH
Europaplatz 3
69115 Heidelberg, Germany

Batch number: 10614746

Printed by Printforce, the Netherlands